# Structured questions in chemistry

# Structured questions in chemistry

## A. L. Barker
Head of Science
Reigate Grammar School

## K. A. Knapp
Formerly of Reigate Grammar School

Macmillan Education

© A. L. Barker & K. A. Knapp 1979

All rights reserved. No part of this
publication may be reproduced or transmitted,
in any form or by any means, without
permission.

First published 1979

Published by
**Macmillan Education Limited**
Houndmills, Basingstoke Hampshire RG21 2XS
and London
Associated companies in Delhi, Dublin,
Hong Kong, Johannesburg, Lagos, Melbourne,
New York, Singapore, and Tokyo

Printed in Hong Kong

*British Library Cataloguing in Publication Data*
Barker, Alan Lewis
   Structured questions in chemistry
   1. Chemistry – Examinations, questions, etc
   I. Title II. Knapp, K.A.
   540'.76  QD42
ISBN 0-333-25663-8

# Contents

*Preface* vi
*A note to the pupil* viii
*Data sheet* ix

## Part 1

1 Separations and purity 1
2 Classes of substance and types of change 2
3 The atmosphere and combustion 3
4 Water and hydrogen 5
5 Acids, bases and salts 6
6 Atoms and molecules 8
7 The periodic table 10
8 Bonding and structure, redox reactions 12
9 Moles, formulae and equations 13
10 The molecular theory of gases 15
11 Electrochemistry 17
12 Solubility 21
13 Rate of reaction 23
14 Reversible reactions 25
15 Energy changes in chemistry 28
16 Carbon 30
17 Nitrogen and phosphorus 31
18 Oxygen and sulphur 33
19 The halogens 35
20 The metals 37
21 Organic chemistry 39
22 Chemical analysis 41

## Part II

General questions 46

Answers to calculations 60

# Preface

This book has been designed for pupils following chemistry courses leading to O Level and similar examinations. It forms an ideal partner for the authors' *Chemistry—a practical approach* (Macmillan Education, 1978) but may, of course, be used on its own.

Structured questions often contain more than one topic, thereby limiting their use. We have tried to overcome this problem by dividing the book into two parts. Each question in Part 1 concerns a single topic, although unifying concepts such as the mole and the periodic table may appear also. The order of the topics given in the contents list is the same as that of the chapter headings in *Chemistry—a practical approach*. In Part II the questions are of a more varied nature, the main topics in each one being indicated in brackets after the number of the question and the order being roughly the same as in Part I. We hope that this splitting of the book into two parts will make it more flexible in use in tests, in class and for homework.

In order to make location of the relevant information easier for the less able pupils some of the questions have been divided into sections. However, in the majority of questions all the information is given at the beginning, since this is the format most commonly encountered in examinations.

Some of the questions largely test factual recall but in the book as a whole we have tried to test a reasonable balance of knowledge, comprehension and higher thought processes, such as the application of principles to unfamiliar situations. Structured questions are particularly suitable for this last category since they lead pupils in steps to answers which they might otherwise be unable to obtain. As the classification of abilities being tested is open to argument and depends on the background of the pupils we have not given a detailed analysis. However, we have included a suggested mark scheme in order to indicate to pupils the depth of answer expected. Totals of 10, 12, 15 and 20 have been given rather than one fixed figure so that each answer may be awarded the mark which in our opinion it merits, when set against the rest of the book. No doubt some teachers will disagree with our suggested mark allocation and will wish to alter it. A data sheet is included on page ix and, as an aid to pupils, questions requiring reference to this have been marked with a '*D*'.

The questions have been pretested at Reigate Grammar School and selections have been used at Varndean Sixth Form College, Brighton, and at Imberhorne School, East Grinstead. Our thanks are due to David Watkins, Peter Crees, Colin Woolcock and the pupils of the three schools for their help in this. As a result of the pretesting, questions which have proved to be more difficult have been marked with a star (★), although this classification can only be used as a general guide since the degree of difficulty of a topic depends to a large extent on the individual teacher's approach. We hope that the pretesting of the material has removed ambiguities and other faults but we would welcome any suggestions for improvements.

The nomenclature used is that recommended by the examining boards, itself largely based on *Chemical Nomenclature, Symbols and Terminology* published by the Association for Science Education (1972).

Our aim has been to write a set of questions in which the less able pupils can obtain reasonable marks and so be encouraged, while at the same time providing sufficient material of a more difficult nature to stimulate the more able. We hope that we have achieved this aim.

A.L.B.
K.A.K.
May 1979

# A note to the pupil

The questions in this book have been designed to test both your memory and your understanding.

Do not give up if you are asked about substances which you have not studied. If you work carefully through the separate parts of the questions you will often find that you can reach the final answers even though the compounds mentioned are not familiar to you.

A '$D$' in the margin means that you need some of the data on page ix in order to complete a calculation. If there is no '$D$' the calculation does not require the data. Stars are used to indicate the more difficult questions.

Answers to the calculations are given on page 60. These should not be consulted until you have finished the questions! Remember that since you have been given the answers your teacher will probably not award you any marks unless you show your working in full.

# Data sheet

**Relative atomic masses**

| Element | Symbol | Relative atomic mass | Element | Symbol | Relative atomic mass |
|---|---|---|---|---|---|
| Aluminium | Al | 27 | Lithium | Li | 7 |
| Barium | Ba | 137 | Magnesium | Mg | 24 |
| Bromine | Br | 80 | Nickel | Ni | 59 |
| Carbon | C | 12 | Nitrogen | N | 14 |
| Chlorine | Cl | 35.5 | Oxygen | O | 16 |
| Chromium | Cr | 52 | Potassium | K | 39 |
| Copper | Cu | 63.5 | Silicon | Si | 28 |
| Hydrogen | H | 1 | Silver | Ag | 108 |
| Iodine | I | 127 | Sodium | Na | 23 |
| Iron | Fe | 56 | Sulphur | S | 32 |
| Lead | Pb | 207 | Zinc | Zn | 65 |

Specific heat capacity of water = $4.2 \text{ J g}^{-1} \text{ K}^{-1}$*.
Density of water = $1.0 \text{ g cm}^{-3}$.

Faraday constant = $96\,500 \text{ C mol}^{-1}$.

Molar volume of gas at s.t.p. = $22.4 \text{ dm}^3 \text{ mol}^{-1}$.
Molar volume of gas at room temperature and pressure = $24 \text{ dm}^3 \text{ mol}^{-1}$.

*The minus sign in the index means 'per'. Thus $4.2 \text{ J g}^{-1} \text{ K}^{-1}$ means 4.2 joules per gram per Kelvin and $1.0 \text{ g cm}^{-3}$ means 1.0 gram per cubic centimetre.

# Part I

### 1  Separations and purity

1.1 A mixture containing powdered samples of ammonium chloride, lead(II) chloride and sodium chloride must be separated into its three constituents. Ammonium chloride is known to sublime; all three substances are soluble in hot water but only ammonium chloride and sodium chloride are soluble in cold water.

  (a) What is meant by sublimation? (2)
  (b) Describe a method by which the ammonium chloride can be separated from the mixture. (3)

  The lead(II) chloride can be separated from the mixture of sodium chloride and lead(II) chloride by the addition of cold water, stirring and filtering.

  (c) Which substance remains in the filter paper? (1)
  (d) How could this substance be obtained as pure, dry *crystals*? (3)
  (e) Name the filtrate. (1)

  The solute may be obtained from the filtrate by evaporating the water. The final evaporation is best done on a steam bath.

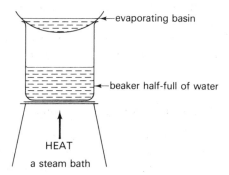

a steam bath

  (f) Explain why the final evaporation is done in this way. (1)
  (g) Suppose that you wanted to obtain pure water from the filtrate instead of the solid dissolved in it. Name the process involved and draw a diagram of the apparatus you would use. (4)

  Total 15

1.2 Small spots of solutions of four yellow compounds, A, B, C and D were placed on a piece of filter paper as shown in the diagram. The spots were allowed to dry and then the filter paper was lowered into a tank containing

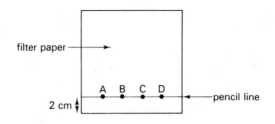

1 cm depth of solvent. The tank was covered and left until the solvent had moved up the filter paper to a distance of about 2 cm from the top. The paper was then removed and the position of the solvent front marked. The distances moved by the solvent and by the spots were then measured.

| Substance | solvent | A | B | C | D |
|---|---|---|---|---|---|
| Distance moved/cm | 11.8 | 5.9 | 4.7 | 2.4 | 8.3 |

(a) What is the name given to this technique? (1)
(b) Calculate the $R_f$ values for (i) A, (ii) B, (iii) C and (iv) D. ($R_f$ value = distance moved by substance/distance moved by solvent front.) (4)
(c) Which of the four substances is probably the most soluble in the solvent? Explain your answer. (2)
(d) Which substance do you think has the greatest tendency to stick to the paper? Explain your reasoning. (2)
(e) An attempt was made to prepare A from D. The product was analysed as above and three spots were obtained. The distance moved by the solvent front was 10.3 cm and the distances moved by the spots were 5.3 cm, 4.0 cm and 2.0 cm. Calculate the $R_f$ values for the spots. (3)
(f) Had all of D reacted? How can you tell? (1)
(g) Was any product other than A obtained? If so, what was it? (2)

Total 15

## 2 Classes of substance and types of change

2.1 Iron and sulphur are two elements. Iron is an example of a metal while sulphur is a non-metal. If a small portion of a mixture of iron filings and flowers of sulphur is heated in an ignition tube, a glow is seen to spread through the mixture and this continues even when the tube is removed from the flame. A black solid remains in the tube at the end.

(a) Explain what is meant by (i) an element, (ii) a mixture. (4)
(b) Give one example of a metal, apart from iron. (1)
(c) State one typical property of a metal. (1)
(d) Give one example of a non-metal other than sulphur. (1)
(e) Give one characteristic property of a non-metal. (1)
(f) How could you separate a sample of iron filings from the mixture? (1)

(g) On heating the mixture, what type of change has taken place, chemical or physical? (1)
(h) What evidence is there to support your answer to (g)? (3)
(i) Name the black solid. Is it an element, a compound or a mixture? (1)
(j) Would it be easy to separate either the iron or the sulphur from the black solid? (1)

Total  15

## 3  The atmosphere and combustion

3.1 Substances need air in order to burn. When magnesium burns in air, the ash left at the end weighs more than the magnesium did. When coal burns, the ash weighs less than the coal.

(a) Why is air necessary for burning to take place? (2)
(b) Why does 'magnesium ash' weigh more than the magnesium? (1)
(c) Why does the ash from coal weigh less than the coal? (1)
(d) Name two gases which pollute the atmosphere in large cities. (2)
(e) Where do these gases come from? (2)
(f) Which two gases do we breathe out in larger quantities than are present in the air we breathe in? (2)
(g) How could you show that these two gases are present in your breath? (3)
(h) Which gas do we breathe out in a smaller quantity than is present in the air we breathe in? (1)
(i) What process replaces this gas in the atmosphere? (1)
(j) Briefly explain this process, mentioning the starting materials, catalyst, source of energy and the products. (5)

Total  20

3.2

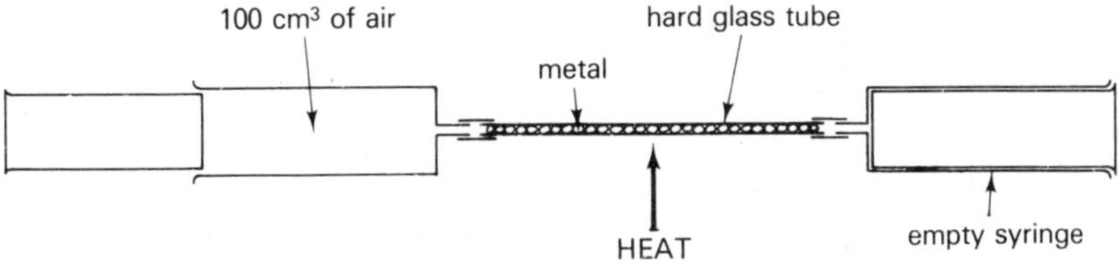

The apparatus shown above is used to determine the percentage of oxygen in the air. The metal is heated strongly and the air passed over it from one syringe to the other. When the tube is cool the volume of the residual gas is noted to be 87 cm$^3$.

(a) Name a suitable metal for this experiment. (1)
(b) What colour change occurs when this metal combines with oxygen? (1)
(c) What volume of the air sample has combined with the heated metal? (1)
(d) Hence, calculate the percentage of the air which has gone. (1)

The metal is heated again and the residual gas passed over it several times until there is no further change in the volume of gas. The metal is found to change colour and after cooling, the volume of residual gas is noted to be 79 cm³.

(e) Why is the volume of residual gas less than in the first experiment? (1)
(f) What is the percentage of oxygen in the air according to this experiment? (1)
(g) Name two processes occurring in everyday life which use up oxygen. (2)
(h) Outline very briefly why it is that the percentage of oxygen in the atmosphere remains approximately constant. (2)

Total 10

3.3

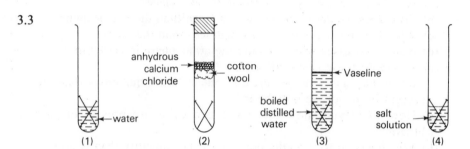

Four tubes were set up as shown to investigate the rusting of iron nails. The Vaseline was placed in tube (3) while the water was still hot. After a few days the nails in tube (4) were very rusty, those in tube (1) were slightly rusty but the ones in tubes (2) and (3) showed no signs of rust.

(a) What is the purpose of the anhydrous calcium chloride in tube (2)? (1)
(b) Why is *boiled* distilled water used in tube (3)? (1)
(c) What is the purpose of the Vaseline in tube (3)? (1)
(d) From the reaction, or lack of it, in tubes (1), (2) and (3), what can you deduce about the conditions needed for rusting? (3)
(e) Why is the result in tube (4) of particular importance to motorists? (2)

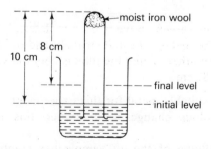

Some moist iron wool was placed in a test tube and the tube was inverted and placed in a beaker of water. The apparatus was inspected each day for one week. The iron rusted and the level of water in the tube rose during the first few days. After this, no further change took place, even though some air remained and not all of the iron was rusty. The air left in the tube was found to put out a burning splint.

(f) What fraction of the air, approximately, has been used up? (1)
(g) Does this experiment show that the air is made up of (i) one, (ii) two, (iii) at least two, gases? Explain your answer. (2)
(h) Which gas has been used up during rusting? How can you tell? (2)
(i) What would be the effect on the level of the water if a larger piece of iron wool were used? How did you reach this conclusion? (2)

Total 15

## 4 Water and hydrogen

4.1 When an electric current is passed through water containing a little sodium fluoride, hydrogen is produced at one electrode and oxygen at the other. The volumes of the gases are in the ratio 2:1.

(a) What is the name given to this process? (1)
(b) What is the purpose of the sodium fluoride? (1)
(c) How could you test for the two gases? (2)
(d) Which gas is given at the positive electrode and which at the negative electrode? (2)
(e) Name a suitable substance for the electrodes. (1)
(f) Draw a diagram of the apparatus used, showing the collection of the gases. (3)
(g) How do you know that the hydrogen and oxygen come from the water? (1)
(h) Does this experiment prove that water is a compound of hydrogen and oxygen only? Explain. (2)
(i) If water *is* a compound of hydrogen and oxygen only, does this experiment *prove* that its formula is $H_2O$? How did you reach your conclusion? (2)

Total 15

4.2
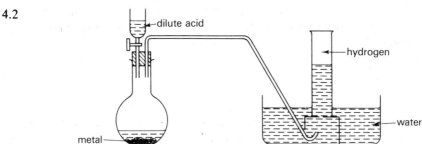

Hydrogen may be prepared by the action of certain dilute acids on certain metals using the apparatus shown. The first few bubbles of gas are allowed to escape before the gas jar is placed in position.

(a) Name a metal and a dilute acid which would be suitable and name
 the product formed in addition to hydrogen. (3)
(b) Write an equation for the reaction. (1)
(c) Name (i) a metal which never gives hydrogen with dilute acids, (ii)
 a dilute acid which normally does not give hydrogen with metals. (2)
(d) Why is the gas jar not placed in position at the beginning of the
 experiment? (1)
(e) Dry hydrogen can be used to reduce some metal oxides. Name a
 metal oxide which could be reduced by hydrogen. (1)
(f) Write an equation for the reaction. (1)
(g) Draw a diagram (excluding the generating flask) showing how the
 experiment may be performed. (3)
(h) When a jet of hydrogen is lit a certain precaution is necessary.
 What is the precaution and why is it necessary? (3)
Total 15

## 5 Acids, bases and salts

5.1 The following instructions refer to the preparation of copper(II) sulphate from dilute sulphuric acid and copper(II) oxide:

'Place 30 cm$^3$ of dilute sulphuric acid in a 100 cm$^3$ beaker. Heat, but do not boil, and add copper(II) oxide in small portions, stirring after each addition until excess solid remains undissolved in the bottom of the beaker. Filter off the solid and heat the filtrate to dryness to obtain crystals of copper(II) sulphate.'

(a) Write an equation for the reaction. (1)
(b) Why is the sulphuric acid heated? (1)
(c) Why is it necessary for excess copper(II) oxide to remain at the
 bottom of the beaker? (1)
(d) Why is the excess copper(II) oxide filtered off? (1)
(e) Heating the filtrate to dryness would not yield *pure crystals* of
 copper(II) sulphate. Why is this? (1)
(f) Describe in detail the method which you would use to obtain dry
 crystals of copper(II) sulphate from copper(II) sulphate solution. (4)
(g) Do you think that the yield of crystals produced is the maximum
 obtainable from this solution? Explain your answer. (1)
Total 10

5.2★ When hydrogen bromide gas, HBr, dissolved in water heat was evolved; when the gas dissolved in methylbenzene there was no noticeable heat change. The solution in water conducted electricity well, giving hydrogen at the cathode. The solution in methylbenzene was a non-conductor. When dry universal indicator papers were dipped into the two solutions the aqueous solution turned the paper red but the methylbenzene solution had no effect.

(a) What type of bonding is present in hydrogen bromide gas? (1)
(b) What does the heat change, or lack of it, tell you about the dissolving of the gas in the two liquids? (2)
(c) Why does the solution in methylbenzene not conduct electricity? (1)
(d) What type of particles must be present in the aqueous solution? (1)
(e) Is the aqueous solution acidic, alkaline or neutral? (1)
(f) What would happen if a *moist* universal indicator paper were placed in the second solution? Explain your answer. (3)
(g) What would you expect to happen if a piece of magnesium ribbon were placed in each solution? (2)

Methylamine gas, $CH_3NH_2$, was substituted for the hydrogen bromide and the whole experiment repeated, using fresh samples of water and methylbenzene. Similar results were obtained but this time less heat was evolved when the gas dissolved in water, the solution was a poor conductor of electricity and it turned universal indicator paper purple. The solution in methylbenzene behaved exactly as before.

(h) What type of bonding is present in methylamine gas? (1)
(i) Is the aqueous solution acidic, alkaline or neutral? (1)
(j) Why is the aqueous solution a much poorer conductor of electricity this time? Is there any evidence to support your answer? (2)

Total 15

5.3 Sodium hydrogensulphate, $NaHSO_4$, is an acid salt. Its solution turns universal indicator paper red.

(a) What is meant by an 'acid salt'? (1)
(b) What, roughly, is the pH of sodium hydrogensulphate solution? (1)

Sodium hydrogencarbonate solution turns universal indicator paper green.

(c) What is the approximate pH of sodium hydrogencarbonate solution? (1)
(d) Is sodium hydrogencarbonate an acid salt? Explain your answer. (2)
(e) What would you expect to see if solutions of sodium hydrogensulphate and sodium hydrogencarbonate were mixed? (1)
(f) Write an equation for the reaction. (2)

Ionic equations for the reactions which occur when the two salts dissolve in water are

$$HSO_4^-(aq) + H_2O(l) \rightleftharpoons H_3O^+(aq) + SO_4^{2-}(aq) \quad \text{(i)}$$
$$HCO_3^-(aq) + H_2O(l) \rightleftharpoons H_2CO_3(aq) + OH^-(aq) \quad \text{(ii)}$$

(g) Define the terms 'acid' and 'base' according to the Brønsted–Lowry Theory. (2)
(h) Write down the two conjugate acid–base pairs in equation (i), making it clear which member of each pair is the acid and which the base. (2)

(i) Do the same for equation (ii). (2)
(j) Which word describes the behaviour of the water in these two reactions? (1)

Total 15

5.4★ The chemistry of liquid ammonia in many respects resembles that of water. Thus water gives rise to the $H_3O^+$ ion in acids such as hydrochloric acid and the $OH^-$ ion in bases such as sodium hydroxide solution,
i.e. $\qquad H_2O(l) + H^+(aq) \rightarrow H_3O^+(aq)$
and $\qquad H_2O(l) - H^+(aq) \rightarrow OH^-(aq)$
Similarly, ammonia yields the $NH_4^+$ ion and the $NH_2^-$ ion.

(a) What is the equivalent of the $H_3O^+$ ion in the ammonia system? (1)
(b) What is the ammonia equivalent of the $OH^-$ ion? (1)
(c) Name the products formed when sodium reacts with water. (2)
(d) Which gas would you expect to be evolved when sodium reacts with liquid ammonia? (1)
(e) Write the formula of the other product of the reaction. (1)
(f) What are the products obtained when hydrochloric acid reacts with sodium hydroxide solution? (2)
(g) What would be the equivalent of hydrochloric acid in the ammonia system? (1)
(h) Write the formula for the ammonia equivalent of sodium hydroxide solution. (1)
(i) Write an equation for the reaction that occurs when the substances mentioned in (g) and (h) are mixed together. (2)

Total 12

## 6 Atoms and molecules

6.1 A solution of $0.1 \text{ cm}^3$ of oil in $1 \text{ dm}^3$ of petroleum ether was made up. A loop of cotton was floated on water and drops of the oil solution were allowed to fall on the surface of the water inside the loop. The loop filled out and became circular and after 3 drops of the oil solution had been added it no longer dented when touched. Its diameter was 14 cm.

In a separate experiment it was found that 20 drops of the oil solution had a volume of exactly $1 \text{ cm}^3$.

(a) When the oil solution was added to the water the petroleum ether evaporated. What happened to the oil? (1)
(b) Why did the loop no longer dent after 3 drops of oil solution had been added? (1)
(c) What was the volume of 1 drop of oil solution? (1)
(d) What volume of oil was present in $1 \text{ cm}^3$ of solution? (1)
(e) What volume of oil was present in 1 drop of solution? (1)
(f) What volume of oil was present inside the loop? (1)
(g) What was the area of the loop? (2)

(h) What was the approximate thickness of the oil film inside the loop? (2)
(i) How is the thickness of the oil film related to the diameter of an oil molecule? What assumption must be made in answering this question? (2)
(j) How could you tell if insufficient oil had been added? (1)
(k) Would it be easy to tell if too much oil had been added? Explain. (2)

Total 15

6.2★

|  | Solid | Liquid | Gas |
| --- | --- | --- | --- |
| Molecular packing | Close | Close | Sparse |
| Molecular movement | Particles vibrate and rotate about fixed points | Particles free to move | Particles free to move |
| Attractive forces | Relatively high | Relatively high | Very small |

Use the information in the table to explain the following observations. In some cases two of the points in the table will be needed.

(a) Solids are rigid but liquids flow. (1)
(b) Gases are easy to compress but liquids are not. (1)
(c) Gases fill any container in which they are placed. (2)
(d) If a jar of bromine vapour and one of air are placed together the brown colour of the bromine soon fills both jars. If a solution of bromine in tetrachloromethane is placed in a test tube with some water, the brown colour only spreads slowly into the water. (4)
(e) The energy required to melt 1 g of ice at 0°C is much less than that required to vaporise 1 g of water at 100°C. (2)

Total 10

6.3 The table gives some information concerning the structures of four atoms, W, X, Y and Z. Work out the missing figures (a) to (l).

|  | Atomic number | Mass number | Number of protons | Number of neutrons | Electronic configuration |
| --- | --- | --- | --- | --- | --- |
| W | 19 | 39 | (a) | (b) | (c) |
| X | (d) | 20 | 10 | (e) | (f) |
| Y | (g) | (h) | 6 | 6 | (i) |
| Z | (j) | (k) | (l) | 8 | 2.4 |

(1) each

Use the letters W, X, Y, Z when answering the following questions.

(m) Which is the atom of a noble gas? (1)
(n) Which two atoms are isotopes of the same element? (1)
(o) Which is a metal atom? (1)

Total 15

6.4 Naturally occurring potassium is a mixture of three isotopes of mass numbers 39, 40 and 41 respectively. The atomic number of each isotope is 19.

(a) Explain what is meant by (i) mass number, (ii) atomic number. (2)
(b) How many (i) protons, (ii) neutrons, (iii) electrons does each isotope contain? (3)
(c) What do you understand by the term 'isotope'? (2)
(d) By means of a diagram show how the protons, neutrons and electrons are arranged in an atom of $^{40}_{19}K$. (2)

A sample of potassium is found to have the composition $^{39}_{19}K$ 93.1%, $^{40}_{19}K$ 0.01%, $^{41}_{19}K$ 6.9%.

(e) Give a definition of relative atomic mass. (2)
(f) Using the above data, calculate the relative atomic mass of potassium. (2)

The isotope $^{40}_{19}K$ is radioactive and has a half-life of approximately $1.4 \times 10^9$ years.

(g) Name the two particles that can be emitted when a nucleus splits up by a radioactive decay process. (2)
(h) What are the relative masses and charges of these two particles? (2)
(i) What is meant by the half-life of an isotope? (1)
(j) If you started with 1 g of $^{40}_{19}K$, how long would it take this mass to be reduced to (i) 0.5 g, (ii) 0.25 g? (2)

Total 20

## 7 The periodic table

7.1 Three elements, X, Y, Z, are in the same period of the periodic table. The table below gives some data concerning the elements and their oxides. One of the elements forms another oxide in addition to that listed.

| Element | X | Y | Z |
| --- | --- | --- | --- |
| Appearance | Shiny black solid | Silvery solid | Yellow crystals |
| Oxide | White solid, $XO_2$ | White solid, YO | White solid, $ZO_3$ |

Use the letters, X, Y, Z, in answering the following questions. Do not try to identify the elements.

(a) What is the oxidation number (or valency) of each element in its oxide? (3)
(b) Write the letters X, Y and Z in the order in which the elements appear in the period. (2)
(c) To which groups do the elements belong? (3)
(d) Which of the elements would you expect to be the best conductor of electricity? (1)

(e) Write the formulae of the compounds which the elements would form with hydrogen. (3)
(f) Which element would combine most readily with chlorine? (1)
(g) Which element would have the lowest melting point? (1)
(h) Which oxide is most likely to be ionic? (1)

Total 15

7.2 Small quantities of oxides of various elements in period 3 are added to distilled water. If the oxide dissolves, the solution is tested with universal indicator. Samples of some of the metallic oxides are then separately boiled with dilute nitric acid and with sodium hydroxide solution. The results of all the experiments are summarised in the table.

| Oxide | $Na_2O$ | $MgO$ | $Al_2O_3$ | $P_2O_5$ | $SO_3$ |
| --- | --- | --- | --- | --- | --- |
| With water | Violent reaction | Almost insoluble | Insoluble | Violent reaction | Violent reaction |
| pH of aqueous solution | 12 | 9 | — | 2 | 2 |
| With nitric acid | Not tested | Dissolves | Dissolves | Not tested | Not tested |
| With sodium hydroxide solution | Not tested | Almost insoluble | Dissolves | Not tested | Not tested |

(a) How does (i) the basic nature, (ii) the acidic nature of the oxides vary across the period? (1)
(b) What connection is there between the properties of an oxide and the metallic or non-metallic nature of the elements in this period? (1)
(c) What trend would you expect to find in the type of bonding in the oxides going across the period? (1)
(d) Which, if any, of the metallic oxides is shown to be (i) acidic only, (ii) basic only, (iii) amphoteric? Explain how you reached your conclusions. (6)
(e) Why would the results obtained by treating sodium oxide with dilute nitric acid and with sodium hydroxide solution be inconclusive? (2)
(f) Silicon, Si, comes between aluminium and phosphorus in period 3. Predict the formula of the oxide of silicon. (1)
(g) What, if anything, can you predict about the oxide of silicon concerning (i) its solubility in water, (ii) its acidic or basic nature? Give reasons for your answers. (4)
(h) Chlorine follows sulphur in period 3 and forms several oxides, all of which are explosive. What is the most likely formula for an oxide of chlorine, following the trend shown in the table? (1)
(i) Is the oxide likely to be basic, amphoteric or acidic? (1)
(j) Why is it easier to predict for chlorine than for silicon? (2)

Total 20

7.3 The data given in the table refer to four elements in the same group of the periodic table.

| Element | Atomic number | Density/g cm$^{-3}$ | Melting point/K (°C) |
|---------|---------------|---------------------|----------------------|
| A | 56 | 3.51 | 987 (714) |
| B | 20 | 1.54 | 1123 (850) |
| C | 12 | 1.74 | 923 (650) |
| D | 38 | 2.62 | 1041 (768) |

(a) Rewrite the table so that the elements are in order of increasing atomic number. (1)
(b) To which group in the periodic table do these elements belong? How did you reach your conclusion? (2)
(c) What is the atomic number of the first member of this group in the periodic table? (1)
(d) What general trend is seen in (i) densities, (ii) melting points going down the group? (2)
(e) What is the oxidation number (or valency) shown by these elements? (1)
(f) Which element would be the most reactive, A, B, C or D? Why is this? (3)

Total 10

## 8 Bonding and structure, redox reactions

8.1 The table gives some properties of three substances, A, B and C.

|  | A | B | C |
|---|---|---|---|
| Appearance | yellow solid | yellow solid | yellow solid |
| m.p./K (°C) | 1336 (1063) | 386 (113) | 675 (402) |
| Solubility in water | insoluble | insoluble | sparingly soluble |
| Electrical conductivity (solid) | conducts | does not conduct | does not conduct |
| Electrical conductivity (molten) | conducts | does not conduct | conducts but is decomposed |

(a) Fit each substance into one of the following structural types, giving your reasons: (i) giant ionic lattice, (ii) giant atomic lattice, (iii) molecular lattice, (iv) giant metallic lattice. (6)
(b) Why does C conduct electricity when molten but not when solid? (1)
(c) Why is A not decomposed when it conducts electricity? (1)
(d) Explain in terms of structure and bonding why the melting point of B is lower than that of C. (3)

(e) Which of these substances, if any, would you expect to dissolve in the organic solvent, methylbenzene? (1)
(f) When molten C conducts electricity a purple vapour is given off at the anode and a bead of molten metal collects under the cathode. Suggest what C might be, giving your reasons. (4)
(g) Suggest, with reasons, what A and B might be. (4)

Total 20

8.2 Consider the reaction between carbon monoxide and iron(III) oxide.

$$3CO(g) + Fe_2O_3(s) \rightleftharpoons 2Fe(s) + 3CO_2(g)$$

(a) What is the oxidation state of carbon (i) in carbon monoxide, (ii) in carbon dioxide? (2)
(b) What is the oxidation state of iron (i) in iron(III) oxide, (ii) in elemental iron? (2)
(c) Give definitions of oxidation in terms of (i) loss or gain of oxygen, (ii) loss or gain of electrons, (iii) change of oxidation number. (3)
(d) Give definitions of reduction in the same ways as (c). (3)
(e) Which of the substances in the reaction is oxidised and which is reduced? (2)
(f) Why is the definition given in (c) (ii) not applicable to the above reaction? (2)

When bromine water is added to a solution of iron(II) bromide, the following reaction occurs.

$$2FeBr_2(aq) + Br_2(aq) \rightarrow 2FeBr_3(aq)$$

(g) Write an ionic equation for the reaction. (2)
(h) Explain the reaction in terms of oxidation and reduction. (4)

Total 20

## 9 Moles, formulae and equations

9.1 Hydrogen is passed over a heated sample of an oxide of lead, A, contained in a tube in order to reduce the lead oxide to lead. The mass of the lead oxide at the beginning of the experiment is 4.46 g whilst at the end of the experiment, the mass of lead remaining is 4.14 g.

(a) How many grams of oxygen are contained in the sample of lead oxide? (1)
D (b) How many moles of oxygen atoms is this? (2)
D (c) How many (i) grams, (ii) moles of lead atoms are present in the sample? (3)
(d) Hence, calculate the empirical formula of the lead oxide. (3)

Another oxide of lead, B, has the percentage composition 90.7% Pb, 9.3% O.

D (e) Calculate its empirical formula. (3)

When a sample of B is heated in an ignition tube oxygen is evolved and the oxide, A, is formed. In an experiment to determine the equation for the reaction it is found that 4.11 g of B yields 0.096 g of oxygen.

D (f) How many moles of oxygen molecules are produced? (2)
D (g) How many moles of B are used? (2)
(h) How many moles of B are required to produce 1 mole of oxygen molecules? (2)
(i) Write a balanced equation for the reaction. (2)

Total 20

9.2 Some iodine was added to a weighed sample of zinc powder in a test tube and the combined mass found. Ethanol was added and the zinc and iodine were seen to react. When the reaction had finished all of the iodine had gone but some zinc was left in the bottom of the tube. The ethanol was removed and the zinc washed with fresh ethanol. It was then dried and weighed.

The results were:
Mass of test tube = 14.52 g
Mass of test tube + zinc at beginning = 14.77 g
Mass of test tube + zinc + iodine = 15.27 g
Mass of test tube + zinc at end = 14.64 g

(a) Why was it necessary to find the mass of the zinc and test tube at the end of the experiment? (1)
(b) What happened to the compound formed when the zinc and iodine reacted? (1)
(c) What mass of zinc reacted with the iodine? (1)
D (d) What fraction of a mole of zinc atoms is this? (2)
(e) Does the mass of the test tube have to be known accurately? Explain your answer. (1)
(f) What mass of iodine reacted? (1)
D (g) What fraction of a mole of iodine atoms is this? (2)
(h) What is the empirical formula of the compound formed between zinc and iodine? (2)
(i) Suggest a name for this compound. (1)

Total 12

9.3★ An experiment was carried out to investigate the reaction between aqueous solutions of copper(II) sulphate and ammonia. 3.0 cm³ portions of 1.0 M copper(II) sulphate solution were measured into seven test tubes. A different volume of 0.75 M ammonia solution was added to each tube and the contests were stirred thoroughly. A pale blue precipitate, thought to be copper(II) hydroxide, formed in each case. All of the tubes were centrifuged for one minute and the heights of the precipitates were measured. The results are shown on the next page.

| Volume of ammonia solution/cm³ | 1.0 | 3.0 | 5.0 | 7.0 | 9.0 | 11.0 | 13.0 |
|---|---|---|---|---|---|---|---|
| Height of precipitate/cm | 0.5 | 1.5 | 2.5 | 2.7 | 2.1 | 1.5 | 0.9 |

(a) Why were all of the tubes centrifuged for the same time? (1)
(b) Draw a graph of height of precipitate/cm ($y$ axis) against volume of ammonia solution/cm³ ($x$ axis). Mark the $y$ axis from 0 to 3 cm and the $x$ axis from 0 to 18 cm³. (3)
(c) From the graph predict (i) the maximum height of the precipitate, (ii) the volume of ammonia solution required to produce this precipitate. (2)
(d) How many moles of $Cu^{2+}$ ions are present in 3.0 cm³ of 1.0 M copper(II) sulphate solution? (1)
(e) How many moles of 'ammonium hydroxide' are present in the volume of 0.75 M ammonia solution predicted in (c)? (1)
(f) What was the ratio of moles of $Cu^{2+}$ ions: moles of 'ammonium hydroxide' in the mixture which gave the most precipitate? (1)
(g) Was the precipitate copper(II) hydroxide? Explain your answer. (2)
(h) From the graph, predict the volume of ammonia solution which would have to be added to just dissolve the precipitate. (1)
(i) How many moles of ammonia would be dissolved in this volume of 0.75 M ammonia solution? (1)
(j) The final solution contains ions with the formula $[Cu(NH_3)_n]^{2+}(aq)$. Use your answers to (d) and (i) to work out the value of $n$. (1)
(k) What would be the colour of this final solution? (1)
Total 15

**10 The molecular theory of gases**

10.1 A certain gas, X, has a mass of 0.084 g and a volume of 44.8 cm³ at s.t.p. On analysis, it is found that X is a hydrocarbon containing 85.7% carbon.

(a) What do you understand by the term 's.t.p. (standard temperature and pressure)'? (1)
D (b) What fraction of a mole is 44.8 cm³ of gas at s.t.p.? (2)
(c) What is the relative molecular mass of X? (2)
(d) Which element, apart from carbon, does X contain? (1)
(e) What percentage of this element is present? (1)
D (f) Calculate the empirical formula of X. (2)
D (g) What is the molecular formula of X? (1)
Total 10

10.2 20 cm³ of a gas $X_2$ combines with 40 cm³ of a gas $Y_2$ to give 40 cm³ of a gaseous product. The volumes are measured at room temperature and pressure.

(a) What is the ratio of volumes of $X_2(g)$: $Y_2(g)$: product(g)? (1)
(b) What is the ratio of moles of $X_2(g)$: $Y_2(g)$: product(g)? (1)
(c) The equation must be of the form:
$$xX_2(g) + yY_2(g) \to ?(g)$$
From your answer to (b) deduce the values of x and y and hence write the complete equation in terms of X and Y. (2)

When the gaseous product of the above reaction is passed over heated iron it is found that $1920\,cm^3$ of the gas combines with $3.36\,g$ of iron to form a solid product and $1920\,cm^3$ of $Y_2$ (all volumes are measured at room temperature and pressure).

D (d) How many moles of gas are used and how many moles of $Y_2$ are produced? (2)
D (e) How many moles of iron atoms are used? (1)
(f) What is the ratio of moles of gas: Fe(s): $Y_2(g)$? (1)
(g) Deduce the equation for the reaction. (Write the formulae of the gases in terms of the symbols X and Y.) (2)

Total 10

10.3★ A solid compound contains 75% by mass of aluminium and 25% by mass of carbon. Its relative molecular mass is 144.

D (a) Calculate the empirical formula of the compound. (2)

When the compound reacts with water, aluminium hydroxide and a colourless gas X are produced. $0.1\,g$ of the compound gives $50\,cm^3$ of the gas at room temperature and pressure.

(b) Draw a diagram of the apparatus you could use to carry out this experiment. (2)
D (c) What is the ratio of moles of solid compound: moles of gas X? (2)

The $50\,cm^3$ of colourless gas requires $100\,cm^3$ of oxygen for complete combustion and forms $50\,cm^3$ of another colourless gas Y, all volumes being measured at room temperature and pressure. Traces of a colourless liquid condense inside the apparatus. The new colourless gas turns lime water milky and the colourless liquid turns a blue cobalt(II) chloride paper pink.

(d) What is the new colourless gas Y and what is the colourless liquid? (2)
(e) What is the ratio of volumes of $X(g)$: $O_2(g)$: $Y(g)$? (1)
(f) What is the ratio of moles of $X(g)$: $O_2(g)$: $Y(g)$? (1)
(g) Use your answer to (f) to write the equation for the combustion of X and to work out the molecular formula of X. (3)
(h) Use your answers to (a) and (c), together with the molecular formula of X deduced in (g) to write the equation for the reaction between the original solid compound and water. (2)

Total 15

## 11 Electrochemistry

**11.1★** An aqueous solution of copper(II) sulphate is electrolysed using carbon electrodes. A pinkish-brown deposit appears on the cathode and a colourless gas is liberated at the anode.

(a) Which ions are present in the solution? (2)
(b) Which ions migrate to the cathode and which one is preferentially discharged? (2)
(c) Which ions migrate to the anode and which one is preferentially discharged? (2)
(d) What is the colourless gas? How would you confirm its identity? (2)
(e) Write equations to illustrate what is happening at each electrode. (4)

After some time, the solution is considerably paler in colour. In addition, it is noticed that the pinkish-brown deposit is forming much more slowly and that a colourless gas is being liberated at the cathode.

(f) Explain why the solution becomes paler in colour. (1)
(g) Name the colourless gas evolved at the cathode. How could you identify this gas? (2)
(h) Why does the formation of the pinkish-brown deposit slow down? Why is the colourless gas evolved? (2)
(i) How could you modify the experiment so that the colourless gas is not evolved at the cathode and the pinkish-brown deposit continues to be formed? (1)
(j) When the deposition of the pinkish-brown substance ceases, it is noticed that the rate of liberation of gas at the cathode is twice that at the anode. Explain why the rates of gas liberation differ in this way. (2)

Total 20

**11.2**

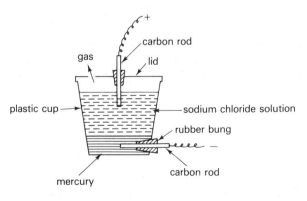

An apparatus for electrolysing sodium chloride solution is made from a plastic cup and set up as shown.*

(a) What is the material of (i) the anode, (ii) the cathode? (2)

*R.W. Chapman, *School Science Review*, **56**, 766 (1975).

The gas escaping from the hole in the lid turns a moist universal indicator paper red and then bleaches it. *No* gas is given off at the cathode.

(b) Name all the ions which are attracted to the anode. Which of these is discharged? (3)
(c) Write an equation showing what happens at the anode. (1)
(d) Name the ions which are attracted to the cathode. Which must be discharged? (3)
(e) If the cathode were made of platinum, which ions would be discharged at it? (1)

The apparatus is dismantled and the mercury is poured into a beaker of water. Bubbles of gas are given off which pop when a flame is applied and the water turns a universal indicator paper purple. The mercury stays in the bottom of the beaker.

(f) What is the colourless gas? (1)
(g) What is present in the aqueous solution? (1)

Total 12

11.3

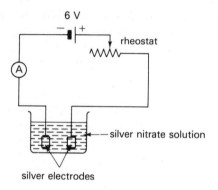

Using the apparatus shown in the diagram, a current of 0.25 A was passed through a solution of silver nitrate. The mass of the cathode was measured at intervals of 20 minutes and the following results were obtained.

| Time/min | 20 | 40 | 60 | 80 | 100 | 120 | 140 | 160 |
|---|---|---|---|---|---|---|---|---|
| Mass of cathode/g | 25.59 | 25.92 | 26.25 | 26.59 | 26.92 | 27.26 | 27.59 | 27.92 |

(a) What is the purpose of the rheostat in the circuit? (1)
(b) Why does the mass of the cathode increase? (1)
(c) Plot a graph of time/min on the $x$ axis against mass of cathode/g on the $y$ axis. Mark the $x$ axis from 0 to 160 min and the $y$ axis from 25.0 to 28.0 g. (2)
(d) What does the shape of the graph tell you about the relationship between time and the increase in mass of the cathode? (1)

18

(e) Do the results agree with Faraday's first law of electrolysis? Explain your answer. (3)
(f) At what point does the graph meet the y axis? (1)
(g) What does this reading represent? (1)
(h) Use the graph to determine what the mass of the cathode was after $2\frac{1}{2}$ hours. (1)
(i) What mass of silver had been deposited after this time? (1)

D (j) How many moles of silver atoms is this? (Answer to 3 decimal places.) (2)
(k) How many coulombs of electricity had passed through the circuit? (2)

D (l) How many moles of electrons is this? (Answer to 3 decimal places.) (2)
(m) What can you deduce from the answers to (j) and (l)? (2)
Total 20

11.4 A current of 0.15 A is passed through two voltameters in series for 45 minutes.

In the first voltameter, containing silver nitrate solution in contact with silver electrodes, 0.453 g of silver is deposited on the cathode.

The second voltameter contains copper electrodes with dilute sulphuric acid. Initially hydrogen is obtained at the cathode but no change is noted at the anode. However, after a short time the solution around the anode becomes noticeably blue in colour.

(a) Write an equation to represent the change taking place at the cathode in the first voltameter. (1)

D (b) How many moles of silver atoms are deposited on the cathode? (2)
(c) How many moles of electrons must flow through the circuit when the number of moles of silver atoms in (b) are deposited? (1)
(d) How many coulombs of electricity flow through the circuit during the experiment? (2)
(e) Calculate the quantity of electric charge which must be carried by 1 mole of electrons. (1)
(f) Hence, what is the value of the Faraday constant? (1)
(g) Write equations to represent the initial changes taking place at the electrodes in the second voltameter. (2)
(h) How many moles of hydrogen atoms will be produced in this experiment? (1)
(i) How many moles of hydrogen molecules is this? (1)

D (j) Hence, what volume of hydrogen, measured at room temperature and pressure, will be obtained? (2)
(k) What might be observed at the cathode if the experiment were allowed to continue for a long time? (1)
Total 15

11.5 Two cells are connected in series, the first containing silver electrodes dipping into silver nitrate solution and the second containing nickel electrodes dipping into nickel sulphate solution. The mass of the silver cathode is 36.21 g and that of the nickel cathode is 28.84 g. A current of $\frac{1}{6}$ A is passed for 3 hr 13 min and the new masses are found to be 38.37 g and 29.43 g respectively.

(a) What should be done to the cathodes before the final weighing? (2)
(b) How many coulombs of electricity were passed? (2)
D (c) How many moles of electrons carry this charge? (2)
(d) What mass of silver was deposited? (1)
D (e) How many moles of silver atoms is this? (2)
(f) Do these results agree with the formula $Ag^+$? (1)
(g) What mass of nickel was deposited? (1)
D (h) How many moles of nickel atoms is this? (2)
(i) Calculate the charge on each nickel ion. (1)
(j) Write down the empirical formula of nickel sulphate. (1)
Total 15

11.6

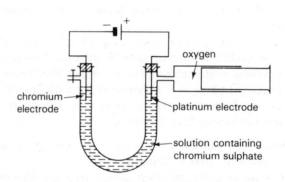

A current was passed through a solution containing chromium sulphate as shown. When 36 cm³ of oxygen (measured at room temperature and pressure) had been collected it was found that the chromium electrode had increased in mass from 2.718 g to 2.822 g.

(a) Which electrode is the cathode and which the anode? (1)
(b) Which ions were attracted to the platinum electrode? (2)
(c) Write an equation for the formation of oxygen at the platinum electrode. (2)
(d) What test could be applied to confirm that the gas was oxygen? (1)
D (e) How many moles of oxygen molecules were formed? (2)
(f) How many moles of electrons passed through the circuit? (2)
D (g) How many coulombs passed? (2)
(h) By how much did the mass of the chromium electrode increase? (1)
D (i) How many moles of chromium atoms were deposited? (2)
(j) How many moles of electrons would deposit 1 mole of chromium atoms? (3)
(k) What is the formula of the chromium ions? (1)
(l) Write down the empirical formula of chromium sulphate. (1)
Total 20

## 12 Solubility

12.1

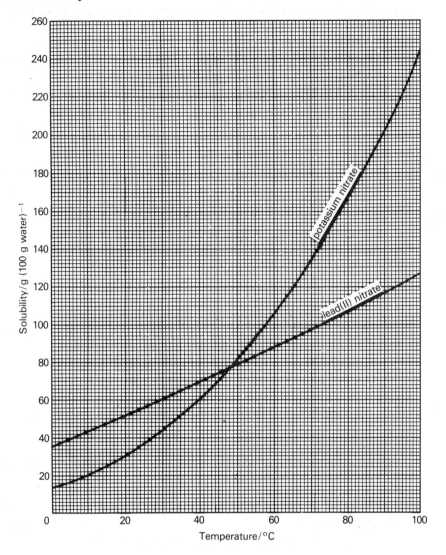

The graph shows the solubilities of lead(II) nitrate and potassium nitrate in water at various temperatures.

(a) What is meant by the term 'solubility'? (1)
(b) What is the solubility of (i) lead(II) nitrate, (ii) potassium nitrate at 100°C? (2)
(c) How much (i) lead(II) nitrate, (ii) potassium nitrate is needed to saturate 100 g of water at 15°C? (2)

A mixture of lead(II) nitrate and potassium nitrate in the proportion of 1:1 by mass is to be separated. 10 g of the mixture is placed in a boiling tube, 10 cm$^3$ of distilled water added and the contents heated to boiling. The liquid is then allowed to cool.

(d) Will both solids dissolve completely at 100°C? (1)
(e) At what temperature will crystals first appear as the solution is cooled? (1)
(f) What will these crystals be? (1)
(g) Would a pure sample of either solid be obtained if the liquid were allowed to cool to 15°C? Explain your answer. (2)
(h) Would a greater degree of purity be obtained if the experiment were repeated using the solid which crystallises out at 15°C and a fresh 10 cm³ portion of distilled water? Explain your reasoning. (2)

Total 12

12.2★ This question concerns four aqueous solutions whose concentrations are as given below:

| | |
|---|---|
| hydrochloric acid | 0.2 mol HCl per dm³ of solution |
| lead(II) nitrate | 66.2 g Pb(NO₃)₂ per dm³ of solution |
| lithium hydroxide | 0.96 g LiOH per 100 cm³ of solution |
| sulphuric acid | 0.2 mol H₂SO₄ per dm³ of solution |

The following data are also needed.

| solute | solubility/g (100 cm³ water)⁻¹ |
|---|---|
| Pb(OH)₂ | 0.016 |
| PbCl₂ | 0.99 |
| LiCl | 83.0 |
| LiNO₃ | 70.0 |
| Li₂SO₄ | 35.0 |

D (a) What is the concentration of (i) lead(II) nitrate solution, (ii) lithium hydroxide solution in mol per dm³? (3)

100 cm³ of lithium hydroxide solution is mixed with 100 cm³ of the hydrochloric acid.

$$LiOH(aq) + HCl(aq) \rightarrow LiCl(aq) + H_2O(l)$$

(b) Is the resulting solution acidic, alkaline or neutral? (1)
D (c) Will sufficient lithium chloride be formed to give a precipitate? (2)
D (d) How, if at all, would the results differ using (i) sulphuric acid, (ii) lead(II) nitrate solution (concentrations as above) instead of the hydrochloric acid? Explain your reasoning. (8)

When 100 cm³ of the lead(II) nitrate solution react with 100 cm³ of hydrochloric acid, the final solution is acidic and a precipitate of lead(II) chloride is formed.

D (e) Explain the observations. (4)
(f) If 200 cm³ of hydrochloric acid were used, what effect would this have on the mass of precipitate formed? Explain your answer. (2)

Total 20

## 13  Rate of reaction

13.1

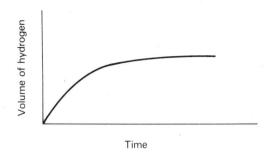

A few pieces of granulated zinc are added to 40 cm³ of 2 M hydrochloric acid in a conical flask and the volume of hydrogen evolved is measured at regular intervals of time in order to obtain the above graph. When the evolution of hydrogen ceases there is still some zinc left in the flask.

(a) Why is all the zinc not used up? (1)
(b) Write the equation for this reaction and suggest a suitable catalyst. (2)
(c) Give two ways in which the reaction could be speeded up other than by the addition of a catalyst. (2)
(d) Draw a graph of rate of the reaction at any instant against time. (2)
(e) The experiment is repeated but this time the mass of the flask is measured at regular intervals. If a graph of decrease in mass of the flask is plotted against time, how would its shape compare with that of the volume against time graph obtained in the first experiment? (1)
(f) How would (i) the average rate, (ii) the total volume of hydrogen evolved, change if the reaction were carried out using 80 cm³ of 1 M hydrochloric acid and the same mass of zinc? (2)
(g) What changes would be observed if the reaction were carried out using 40 cm³ of 3 M hydrochloric acid and the same mass of zinc? (2)
Total 12

13.2  Ten marble chips of fairly uniform size are selected and their reaction with hydrochloric acid studied.

40 cm³ of 2 M hydrochloric acid is placed in a conical flask, one marble chip added and a plug of cotton wool quickly inserted in the neck of the flask. The flask is placed on a balance, the mass noted and a stop clock immediately started. After 1 minute, the mass is again noted, a second marble chip added and the whole procedure repeated. This is continued until all ten chips have been added. The results are shown overleaf.

When all ten chips have been added the mass is noted at 1 minute intervals for a further 30 minutes. After this time, marble chips are still visible in the flask but the mass is not changing.

| Number of marble chips added | Mass loss during 1 minute/g |
|---|---|
| 1 | 0.010 |
| 2 | 0.019 |
| 3 | 0.029 |
| 4 | 0.037 |
| 5 | 0.046 |
| 6 | 0.054 |
| 7 | 0.062 |
| 8 | 0.071 |
| 9 | 0.078 |

(a) Why is a plug of cotton wool placed in the neck of the flask? (1)
(b) Plot a graph of loss in mass during 1 minute (on the $y$ axis) against number of marble chips added (on the $x$ axis). (2)
(c) Predict what the loss in mass will be during the minute after the tenth chip has been added. (2)
(d) Explain why the graph is not a straight line. (2)
(e) Draw a sketch of the graph showing how the decrease in mass varies with time once all the marble chips have been added. (2)
(f) Why is the mass not changing at the end of 30 minutes even though there are still marble chips in the flask? (1)
(g) How would the decrease in mass vary with time if 40 cm³ of 1 M hydrochloric acid were used? (1)
(h) Would marble chips still be visible after 30 minutes if 1 M acid were used? (1)

Total 12

13.3 Nitrogen, boiling point 77 K (−196°C), and hydrogen, boiling point 20 K (−253°C), are combined together to give ammonia, boiling point 240 K (−33°C), according to the equation

$$N_2(g) + 3H_2(g) \rightleftharpoons 2NH_3(g).$$

When the reaction is carried out in a glass vessel the reaction rate is very slow but if an iron vessel is used a much greater rate of reaction is achieved.

(a) Why is the reaction rate greater in the iron vessel? (1)
(b) Give a definition of the term used to describe the function of the iron. (2)
(c) What would happen to the rate of the reaction if some finely powdered iron were added to either vessel? (1)
(d) How would you expect the rate of this reaction to change if the temperature were raised? (1)
(e) Explain your answer to (d). (2)

The table gives the percentages of ammonia in the equilibrium mixture at various temperatures.

| Temperature/K (°C) | Yield of ammonia at 250 atm pressure |
|---|---|
| 1273 (1000) | negligible |
| 823 (550) | 15% |
| 473 (200) | 88% |

In practice, a temperature of 773 K (500°C) is chosen.

(f) Suggest one reason for not operating the plant at a temperature greater than 773 K. (2)
(g) Why is a temperature of much less than 773 K not advisable? (1)
(h) Increasing the pressure raises the yield of ammonia. Give one reason why the industrial process is not usually carried out at a higher pressure than 250 atm. (1)
(i) How could the ammonia be separated from the nitrogen and hydrogen in the equilibrium mixture? (1)

Total 12

## 14 Reversible reactions

14.1 (i) Solutions containing iron(III) ions and thiocyanate ions react to produce a blood-red colour:

$$Fe^{3+}(aq) + CNS^-(aq) \rightleftharpoons [Fe(CNS)]^{2+}(aq)$$
yellow    colourless    blood-red

Provided that the solutions are dilute, the colour of the $Fe^{3+}(aq)$ ion is not noticed.

A mixture of iron(III) chloride and potassium thiocyanate solution containing the above ions in equilibrium is made up so that the colour is pale pink. When one drop of iron(III) chloride solution is added the colour darkens.

(a) In what way does the increase in concentration of one of the reactants on the left hand side of the equation alter the rate of the forward reaction? (1)
(b) How and why does this cause the equilibrium position to alter? (2)
(c) How would you expect the colour to change if 1 drop of potassium thiocyanate solution were added to the original solution? (1)

Fluoride ions can remove $Fe^{3+}(aq)$ ions from solution. The addition of a little solid potassium fluoride to the original pink solution causes the colour to lighten.

(d) In what way does a decrease in concentration of one of the reactants on the left hand side of the equation alter the rate of the forward reaction? (1)
(e) How and why does this cause the equilibrium position to alter? (2)

(ii) When bismuth chloride is dissolved in hydrochloric acid a clear solution is obtained in which the following equilibrium has been established:

$$BiCl_3(aq) + H_2O(l) \rightleftharpoons BiOCl(s) + 2HCl(aq)$$

(f) Is the position of equilibrium well to the left or to the right, in the solution described above? (1)

(g) What would you expect to see on adding water to the solution? Explain your prediction. (2)

(h) How would the addition of concentrated hydrochloric acid to the product in (g) alter the system? What would you observe? (2)

Total 12

14.2★ At room temperature 'nitrogen dioxide' actually consists of a mixture of dinitrogen tetraoxide, $N_2O_4$, and nitrogen dioxide, $NO_2$, in dynamic equilibrium.

$$N_2O_4(g) \rightleftharpoons 2NO_2(g)$$
pale yellow   dark brown

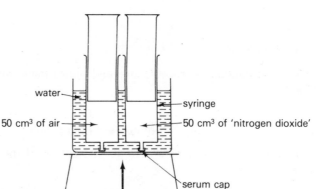

The water in the apparatus shown above is heated until it boils. Both gases expand and the colour of the 'nitrogen dioxide' becomes darker. The final volume of the air is 62 cm³ and that of the 'nitrogen dioxide' is 98 cm³. On cooling to room temperature the two gases contract to their original volumes and the 'nitrogen dioxide' reverts to its original colour.

(a) What is meant by dynamic equilibrium? (2)
(b) What can you say about the original numbers of molecules in the two syringes? Give a reason for your answer. (2)
(c) What must happen to the number of molecules in the 'nitrogen dioxide' sample as it is heated? Explain how you reached your conclusion. (2)
(d) Why does the colour of the 'nitrogen dioxide' darken? (1)

(e) As the temperature rises, what must happen to the relative rates of the forward and backward reactions and to the position of equilibrium? (2)
(f) State Le Chatelier's principle. (2)
(g) Use Le Chatelier's principle to deduce whether the forward reaction in the equation is endothermic or exothermic. Explain your reasoning. (2)
(h) What would you expect to happen to (i) the relative volumes of the gases, (ii) the colour of the 'nitrogen dioxide' if the syringes were cooled to 5°C? (2)

Total 15

14.3★ Read the introduction to the previous question.

The plunger of a sealed gas syringe containing 100 cm³ of 'nitrogen dioxide' at room temperature is pushed in quickly until the volume is 50 cm³. The brown gas immediately *darkens* in colour and then gradually *lightens* until it is paler than it was originally.

(a) Why does the colour of the gas darken at first? (1)
(b) Explain the lightening of colour in terms of the variation in the relative rates of the forward and backward reactions and the position of equilibrium. (2)
(c) Show how Le Chatelier's principle could have been used to predict the results of the experiment. (2)

The plunger of a similar syringe containing 50 cm³ of 'nitrogen dioxide' is quickly pulled out until the volume of gas is 100 cm³.

(d) What would you expect to see at first? Explain your answer. (2)
(e) In what way would the colour of the gas change if the plunger were held at the 100 cm³ mark? (1)
(f) How did you reach your conclusion in (e)? (2)

Total 10

14.4 Ammonia is manufactured by the Haber process, the equation for the reaction being
$$N_2(g) + 3H_2(g) \rightleftharpoons 2NH_3(g).$$

Various experiments are carried out to find the effect of temperature and pressure on the yield of ammonia and the following results are obtained.

| Temperature/K (°C) | Yield of ammonia at 250 atm with a catalyst |
|---|---|
| 1273 (1000) | negligible |
| 823 (550) | 15% |
| 473 (200) | 88% |

| Pressure/atm | Yield of ammonia at 823 K with a catalyst |
|---|---|
| 1 | negligible |
| 100 | 7% |
| 1000 | 41% |

(a) State Le Chatelier's principle. (2)
(b) Which of these temperatures would you choose for the manufacture of ammonia? (1)
(c) Do you think that the production of ammonia is exothermic or endothermic? Explain your answer. (2)
(d) Why do you think that the temperature you have chosen in (b) gives a good yield of ammonia? (2)
(e) Which pressure would you choose for the manufacture of ammonia? Explain why this pressure gives a good yield of product. (2)

In practice, a temperature of 773 K (500°C) and a catalyst are used.

(f) Why is this temperature chosen? (1)
(g) What is the purpose of the catalyst? (1)

Usually, dynamic equilibrium has not been reached by the time the gases leave the catalyst chamber so that only about 10% of the gases have combined.

(h) What is meant by dynamic equilibrium? (2)
(i) Suggest how the process could be modified to obtain dynamic equilibrium. Why is this not done in the industrial process? (2)

Total  15

### 15 Energy changes in chemistry

15.1 The enthalpies of combustion of a number of cycloalkanes are as follows:
cyclopentane $C_5H_{10}$ $\Delta H = -3317$ kJ mol$^{-1}$
cyclohexane $C_6H_{12}$ $\Delta H = -3948$ kJ mol$^{-1}$
cycloheptane $C_7H_{14}$ $\Delta H = -4632$ kJ mol$^{-1}$

When the enthalpy of combustion of cyclononane, $C_9H_{18}$, is measured, the following data is obtained: the combustion of 0.21 g of cyclononane raises the temperature of water in a can (total heat capacity = 0.996 kJ K$^{-1}$) by 10 K.

D (a) Calculate the enthalpy of combustion of cyclononane, using the expression:

enthalpy of combustion/kJ mol$^{-1}$

$= -\left(\dfrac{\text{molar mass of cyclononane}}{\text{mass of cyclononane burned}} \times \text{heat capacity} \times \text{temperature rise}\right)$. (2)

(b) Plot a graph of enthalpy of combustion of the cycloalkanes (on the y axis) against number of carbon atoms in their molecules (on the x axis). (3)

(c) An unknown cycloalkane has an enthalpy of combustion of −5305 kJ mol⁻¹. Use the graph to find how many carbon atoms this cycloalkane contains. (2)
(d) What is the formula of the unknown cycloalkane? (1)
(e) Write an equation to represent its combustion to carbon dioxide and water. (2)

Total 10

15.2★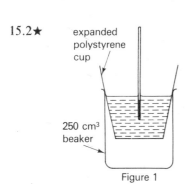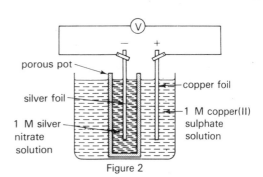

Figure 1     Figure 2

Using the apparatus shown in Figure 1, 0.8 g of powdered copper is added to 100 cm³ of 0.2 M silver nitrate solution and the mixture stirred gently with the thermometer. The temperature at the beginning of the experiment is 18.2°C and it rises to a maximum of 21.7°C. The equation for the reaction is

$$2AgNO_3(aq) + Cu(s) \rightarrow Cu(NO_3)_2(aq) + 2Ag(s).$$

(a) Why is a polystyrene cup used instead of a glass container and why is it placed in the beaker? (2)
(b) How many moles of silver nitrate are present initially? (2)
(c) How many moles of copper atoms are needed to react with the silver nitrate? (2)
D (d) How many moles of copper atoms are present? Is this enough? (2)
(e) What is the temperature rise? (1)
D (f) Calculate the heat produced during the reaction, assuming that the density and specific heat capacity of the solution are the same as those of water. (2)
(g) Calculate the enthalpy of reaction in kJ per mole of copper atoms. (2)

In a second experiment the cell shown in Figure 2 is set up and its e.m.f. is found to be 0.46 V. The ionic equation for the reaction is

$$Cu(s) + 2Ag^+(aq) \rightarrow Cu^{2+}(aq) + 2Ag(s).$$

(h) Write the ionic equation for the reaction taking place in the polystyrene cup in the first experiment. What can you say about the two reactions? (2)
(i) How many moles of electrons must flow through the circuit for 1 mole of copper atoms to go into solution at the anode of the cell and for 2 moles of silver atoms to be deposited on the cathode? (1)
D (j) What is the charge carried by this amount of electrons? (2)

(k) When a cell is working,
number of joules released = potential difference × number of coulombs passed.
Calculate the maximum number of joules which could be released in the above cell when 1 mole of copper atoms dissolve, assuming that the working voltage is the same as the e.m.f. (1)
(l) Does the cell convert all of the energy available from the chemical reaction into electrical energy? (1)

Total 20

15.3★ Using the apparatus shown in Figure 1 of the previous question, 1.6 g of anhydrous copper(II) sulphate is dissolved in 50 cm³ of water, producing a temperature rise of 2.7°C.

D (a) Calculate the heat produced during the dissolving, assuming that the density and specific heat capacity of the solution are the same as those of water. (2)
D (b) How many moles of anhydrous copper(II) sulphate are used (to 2 decimal places)? (1)
(c) Calculate the enthalpy of solution of anhydrous copper(II) sulphate. (2)

A similar experiment is carried out using 2.5 g of copper(II) sulphate-5-water in place of the anhydrous salt, producing a temperature *drop* of 0.5°C.

D (d), (e), (f) as (a), (b), (c) but for the hydrated salt. (2,1,2)
(g) Express the two results on *one* energy level diagram. (4)
(h) Use the diagram to find the enthalpy change for the following reaction:

$$CuSO_4(s) + 5H_2O(l) \rightarrow CuSO_4.5H_2O(s)$$ (2)

In both experiments the final solution contains hydrated copper ions and hydrated sulphate ions.

(i) Is the hydration of ions an endothermic or an exothermic process? (1)
(j) Is the breaking up of a crystal lattice an endothermic or an exothermic process? (1)
(k) Referring to your answers to (i) and (j), explain why the enthalpies of solution of the anhydrous and hydrated forms of copper(II) sulphate differ. (2)

Total 20

## 16 Carbon

16.1 Stalagmites and stalactites grow in underground caves in limestone areas. Chemical analysis shows that they consist of calcium carbonate.
When a small piece of a stalactite is heated strongly it hardly changes in appearance but a colourless gas A is given off which turns lime water milky. A few drops of distilled water are added to the cooled solid residue

B and it becomes very hot and swells up, finally crumbling to a white powder C. The white powder is only slightly soluble in water. An aqueous solution of it turns cloudy and then clear again when breathed into through a drinking straw.

(a) What is the colourless gas A? (1)
(b) Name the solid residue B. (1)
(c) Write an equation for the decomposition of the stalactite on heating. (1)
(d) What is the white powder C? (1)
(e) Why does the solution of C turn cloudy when breathed into? (2)
(f) Write an equation for the reaction. (1)
(g) Why does the solution turn clear again? (1)
(h) What would be seen if the clear solution were boiled for some time? Explain the observations. (2)
(i) Why is this reaction important in the home and in industry? (1)
(j) Explain briefly, without giving equations, how stalactites and stalagmites form. (2)
(k) What would you expect to see if dilute hydrochloric acid were added to a piece of a stalactite? Write an equation for the reaction. (2)

Total 15

16.2 Carbon monoxide can be prepared by passing carbon dioxide from a Kipp's apparatus over strongly heated carbon. The carbon dioxide is first bubbled through water and then through concentrated sulphuric acid. The carbon monoxide is usually passed through concentrated potassium hydroxide solution before being collected.

(a) Why is the carbon dioxide bubbled through (i) water, (ii) concentrated sulphuric acid? (2)
(b) Why is the carbon monoxide passed through concentrated potassium hydroxide solution? (1)
(c) If a flame is applied to a test tube-full of carbon monoxide, would the residual gas give a positive test with lime water? (1)
(d) Write equations for any reactions that occur in (c). (2)
(e) Where might you see the carbon monoxide flame in everyday life? (1)

When carbon monoxide is passed over strongly heated iron(III) oxide, iron is obtained.

(f) What type of reaction occurs? (1)
(g) Write an equation to represent the reaction. (1)
(h) Where is this reaction encountered on an industrial scale? (1)

Total 10

### 17 Nitrogen and phosphorus

17.1 (a) Give two facts which suggest that air is a mixture of gases. (2)

A pupil set up the apparatus shown overleaf in order to prepare a sample of nitrogen.

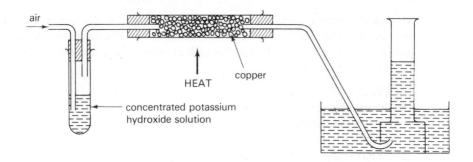

(b) Which gas is removed by the potassium hydroxide solution? (1)
(c) Write an equation to show what happens when this gas reacts with the potassium hydroxide solution. (1)
(d) What would you expect to observe when air is passed over heated copper? (1)
(e) Name the compound of copper produced by the reaction. (1)
(f) The gas which is collected over water is not pure nitrogen. Name two other gases which it contains. (2)
(g) Would you expect the density of this gas to be (i) greater than, (ii) less than, (iii) equal to that of pure nitrogen? (1)
(h) Describe in outline how nitrogen is obtained on an industrial scale. (3)

Total 12

17.2★ Methylamine, $CH_3NH_2$, is a colourless gas with a pungent, choking smell. It is very soluble in water and the resulting solution is alkaline.

(a) What type of bonding must be present in methylamine? (1)
(b) Which characteristic ion must an aqueous solution of methylamine contain? (1)
(c) Which fairly common gas does methylamine resemble? (1)
(d) What would you expect to see if methylamine and hydrogen chloride were mixed? Suggest a formula for the product. (2)
(e) What would be given off if the product in (d) were heated with sodium hydroxide solution? (1)
(f) Write an equation for the reaction in (e). (1)
(g) What would you expect to see if a *little* aqueous methylamine solution were added to copper(II) sulphate solution? Name the coloured product. (2)
(h) What would happen if excess aqueous methylamine solution were added to the mixture in (g)? (1)

Total 10

17.3 When calcium is heated strongly in nitrogen a yellow solid is produced. This solid is called calcium nitride and has the formula $Ca_3N_2$. Calcium nitride reacts with water to form calcium hydroxide and ammonia.

Phosphorus is in the same group of the periodic table as nitrogen (group V). If calcium is heated in phosphorus vapour a red solid is formed. This red solid reacts with water to give a white solid and a colourless gas.

(a) Write an equation for the reaction between calcium nitride and water. (2)
(b) Draw a diagram of an ammonia molecule: only the outermost shell of each atom need be shown. (1)
(c) What type of bonding does this illustrate? (1)
(d) Suggest a suitable name for the red solid. (1)
(e) What is the most likely formula for this solid? (1)
(f) Write an equation for the reaction between the red solid and water. (1)
(g) What would you expect to see if the colourless gas produced in (f) were mixed with hydrogen iodide gas? Suggest a formula for the product. (2)
(h) If this product were heated, would you expect it to (i) melt, (ii) sublime, (iii) decompose? Give a reason for your answer. (2)
(i) What would happen if the product in (g) were heated with sodium hydroxide solution? (2)
(j) Write an equation for the reaction. (2)

Total 15

## 18 Oxygen and sulphur

18.1 Oxygen can be prepared by adding hydrogen peroxide solution to manganese(IV) oxide. The manganese(IV) oxide is recovered unchanged in mass at the end of the reaction.

(a) What is the function of the manganese(IV) oxide in this reaction? (1)
(b) Draw a diagram of the apparatus which you could use to prepare and collect several gas jars of oxygen. (3)

Sodium and sulphur both burn in oxygen and the product of each reaction is soluble in water.

(c) Name the product formed when each element burns in oxygen. (2)
(d) Give the names and approximate pH values of the two aqueous solutions. (3)

When zinc combines with oxygen the compound formed is insoluble in water but dissolves in both dilute hydrochloric acid and in sodium hydroxide solution.

(e) What is the word used to describe the behaviour of the zinc compound? (1)
(f) Write chemical equations for the two reactions of the zinc compound. (2)

Total 12

18.2 Sulphur and selenium are both found in group VI of the periodic table, selenium coming directly below sulphur in the table.

  (a) Would you expect sulphur and selenium to show similar or different chemical properties? (1)
  (b) Which element do you think will be the more reactive as a non-metal? Give a reason for your answer. (3)
  (c) Given iron and selenium, describe how you could prepare a sample of iron(II) selenide. (2)
  (d) Suggest how you might obtain some hydrogen selenide from the iron(II) selenide. (1)
  (e) If hydrogen selenide is passed into an acidified solution of potassium manganate(VII), the solution is decolourised and a precipitate is formed. Name the precipitate. (1)

  Selenium forms a solid dioxide, $SeO_2$, which dissolves in water to give the acid $H_2SeO_3$. This acid forms hydrogenselenites and selenites.

  (f) Give the formulae for sodium hydrogenselenite and sodium selenite. (2)
  (g) Write equations to show how both sodium salts might be formed from the acid. (2)

  Total 12

18.3 Sulphur dioxide may be prepared by warming sodium sulphite crystals with dilute hydrochloric acid.

  (a) Draw a diagram of the apparatus you would use to prepare several gas jars of dry sulphur dioxide. (4)
  (b) How would you tell when each jar was full? (1)

  When aqueous sulphur dioxide solution (sulphurous acid, $H_2SO_3(aq)$) is added to potassium manganate(VII) solution the purple solution is decolourised.

  (c) What does this reaction tell you about sulphurous acid? (1)

  When sulphurous acid is added to an aqueous solution of hydrogen peroxide, $H_2O_2$, there is no *visible* change. However, if dilute hydrochloric acid is added to the mixture, followed by barium chloride solution, a white precipitate is formed.

  (d) What is the white precipitate? Which ion does its formation indicate? (2)
  (e) Explain the reaction between sulphurous acid and hydrogen peroxide solution in terms of oxidation and reduction and write an equation to represent it. (2)

  Magnesium ribbon burns in a jar of sulphur dioxide, producing a white ash and yellow specks of another solid.

  (f) What are the two solids most likely to be? (2)

(g) Explain the reaction in terms of oxidation and reduction and write an equation to represent it. (2)
(h) From your answers to (e) and (g), does sulphur dioxide behave as an oxidising agent, a reducing agent, or both? (1)

Total 15

19.1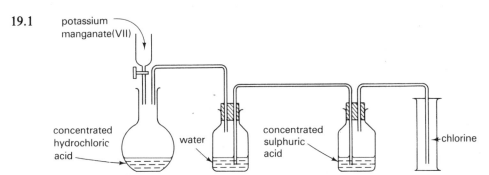

**19 The halogens**

The apparatus shown above is intended to be used for the laboratory preparation of chlorine but contains three mistakes.

(a) What are the mistakes in the diagram? (3)
(b) Why is the gas bubbled through (i) water, (ii) concentrated sulphuric acid? (2)
(c) How can you tell when the gas jar is full? (1)

When a burning taper is plunged into a jar of chlorine, black particles of a solid A and steamy fumes of a gas B are obtained.

(d) Name the solid A and the gas B. (2)
(e) What type of chemical reaction has occurred? (1)
(f) Give one other example of a reaction in which chlorine is behaving in a similar way and state what you would observe. Give an equation for the reaction. (3)

Total 12

19.2 If concentrated sulphuric acid is added to sodium chloride, a colourless gas is evolved which fumes in moist air and turns a moist universal indicator paper red. With sodium bromide the acid gives a similar colourless gas, but a reddish-brown vapour is also produced. If the acid is added to sodium iodide, very little colourless gas is evolved but black crystals are formed and a smell of bad eggs is noticed.

(a) Name the colourless gas in the first experiment. (1)
(b) Name the colourless gas and the reddish-brown vapour in the second experiment. (2)
(c) By what chemical process is the reddish-brown vapour obtained from the colourless gas? (2)

(d) What are the black crystals and the gas which smells of bad eggs? (2)
(e) By what chemical process is the gas which smells of bad eggs obtained from the concentrated sulphuric acid? (2)
(f) Astatine (At) comes below iodine in group VII of the periodic table. Would you expect hydrogen astatide (HAt) to be a good reducing agent or a poor one? Explain your answer. (3)
(g) What would you expect to see if ammonia and hydrogen astatide were mixed? Name the product. (2)
(h) Predict the colour of silver astatide. (1)
(i) How could silver astatide be made? (2)
(j) Write an equation for the reaction described in (i). (1)
(k) Would you expect silver astatide to be soluble in ammonia solution? How did you come to your conclusion? (2)

Total 20

19.3 When chlorine is bubbled into separate aqueous solutions of potassium bromide and potassium iodide, both solutions turn brown. The colours are due to the formation of bromine and iodine respectively.

The table gives some data concerning chlorine, bromine and iodine.

| Property | Chlorine | Bromine | Iodine |
| --- | --- | --- | --- |
| Electronic configuration | 2,8,7 | 2,8,18,7 | 2,8,18,18,7 |
| Atomic radius/nm | 0.099 | 0.114 | 0.133 |
| Ionic radius/nm | 0.181 | 0.195 | 0.216 |

(a) How could the experiment be modified to make the bromine and iodine give distinguishable colours? What would these colours be? (3)
(b) What is the charge on a bromide ion and on a chloride ion? (1)
(c) Why do the ions have these charges? (2)
(d) Why are the ions larger than the corresponding atoms? (1)
(e) Explain the reaction between chlorine and potassium bromide solution in terms of transfer of electrons between atoms and ions. (2)
(f) What type of chemical reaction is this? (1)
(g) What reaction, if any, would you expect to take place between bromine and aqueous solutions of (i) potassium chloride, (ii) potassium iodide? Explain your answer. (4)

Astatine (At) has electronic configuration 2,8,18,32,18,7 and atomic radius 0.140 nm.

(h) What is the formula of potassium astatide? (1)
(i) Predict the radius of the astatide ion. (2)
(j) Would astatide ions be more stable, or less stable, than iodide ions? Explain your answer. (3)

Total 20

## 20 The metals

**20.1** (a) Describe what is seen when a small piece of sodium is placed on the surface of some water. (2)
(b) Write an equation for the reaction that occurs. (1)

The reaction between magnesium and water is very slow but can be speeded up by heating the metal in steam.

(c) Draw a diagram of the apparatus used to investigate this reaction. (2)
(d) Name the products of the reaction. (1)

The reaction between iron and water is also very slow. Again, steam can be used to speed up the reaction but this time the reaction is reversible.

(e) Write an equation for the reaction. (1)
(f) What is meant by a reversible reaction? (1)
(g) Using the apparatus drawn in (c) the reaction goes to completion. Why is this? (2)

Total 10

**20.2**

| Element | Atomic radius/nm | Electronic configuration | Density/g cm$^{-3}$ |
|---|---|---|---|
| Na | 0.157 | 2,8,1 | 0.97 |
| K | 0.203 | 2,8,8,1 | 0.86 |
| Rb | 0.216 | 2,8,18,8,1 | 1.53 |

The table shows some data for the elements sodium (Na), potassium (K) and rubidium (Rb).

(a) How can you tell that rubidium is in the same group of the periodic table as sodium and potassium? (1)
(b) What is the first member of this group in the periodic table? (1)
(c) Would you expect rubidium to be more reactive than potassium, or less reactive? (1)
(d) Why do the reactivities of the two elements differ? (2)
(e) Potassium has a larger atomic radius than sodium and its density is lower. Rubidium has a larger atomic radius than potassium but its density is *higher*. Why should this be? (The method of packing of the atoms is the same for all three.) (3)
(f) Describe what you would expect to see if a small piece of rubidium were added to water. (3)
(g) Write an equation for the reaction. (1)
(h) Would the final solution be acidic, alkaline or neutral? (1)
(i) Write down the formulae of rubidium chloride and rubidium sulphate. (2)
(j) What type of bonding would be present in rubidium chloride? (1)

(k) How would rubidium nitrate behave on heating and what would be formed? (3)
(l) Write an equation for the reaction. (1)

Total 20

20.3 The following experiment is designed to measure the amount of hardness in various samples of water.

50 cm³ portions of the water samples listed in the table are placed in separate conical flasks. Soap solution is added from a burette, the flask is stoppered and then shaken vigorously. More soap solution is added until a permanent lather is obtained.

| Water sample | Volume of soap solution/cm³ | Volume of soap solution to overcome hardness/cm³ |
|---|---|---|
| Distilled water | 0.5 | 0 |
| Tap water | 12.5 | 12.0 |
| Tap water boiled for 15 min and then made up to original volume with distilled water | 9.5 | 9.0 |
| 0.002 M calcium chloride solution | 8.5 | 8.0 |

(a) What is meant by hard water? (1)
(b) There are two types of hardness: temporary hardness and permanent hardness. Explain the difference between the two. (1)
(c) What kind of hardness remains in the boiled water sample? (1)
(d) Hence calculate the percentages of temporary and permanent hardness in the tap water. (2)

It can be shown that calcium ions are the cause of hardness in many water samples.

(e) How many moles of $Ca^{2+}$ ions are present in 50 cm³ of 0.002 M calcium chloride solution? (2)
(f) How many cm³ of soap solution are needed to overcome the hardness caused by the calcium ions in (e)? (1)
(g) How many moles of $Ca^{2+}$ ions are equivalent to 1 cm³ of soap solution? (1)
(h) From your answer to (g), calculate the number of moles of $Ca^{2+}$ ions present in the unboiled sample of tap water, assuming that they are the only metallic ions present. (1)
(i) Suggest two ways that could be used to overcome the hardness remaining in the boiled water sample. (2)

Total 12

20.4 Magnesium, $^{24}_{12}$Mg, and strontium, $^{88}_{38}$Sr, are both in the same group of the periodic table, strontium coming lower down than magnesium.

(a) How many electrons do the atoms of these elements have in their outermost shells? (1)
(b) In which group of the periodic table are these elements found? (1)
(c) What oxidation number (or valency) is shown by these elements in their compounds? (1)
(d) Explain how the reactivity of a metal is related to the size of its atoms. (2)
(e) Which of these two metals is the more reactive? (1)
(f) The ionic radii of these elements are very much smaller than the atomic radii. Why is this? (2)
(g) Write down the electronic configuration of (i) an atom of magnesium, (ii) an ion of strontium. (2)
(h) How many (i) protons, (ii) neutrons would an ion of strontium contain? (2)

Total 12

## 21 Organic chemistry

21.1 Pent-l-ene, $C_5H_{10}$, has the graphic formula

$$\begin{array}{c} \quad\; H \;\; H \;\; H \;\; H \\ \quad\; | \;\;\;\; | \;\;\;\; | \;\;\;\; | \quad\quad\;\; H \\ H-C-C-C-C=C \\ \quad\; | \;\;\;\; | \;\;\;\; | \quad\quad\quad\;\; H \\ \quad\; H \;\; H \;\; H \end{array}$$

When a solution of bromine in tetrachloromethane is added to a few drops of pent-1-ene in a test tube, and the test tube shaken, the bromine solution is decolourised. In a similar experiment, using butter in place of the pent-1-ene, no change is observed. However, when soft margarine replaces the pent-1-ene, decolourisation of the bromine solution occurs as before.

(a) To which homologous series does pent-1-ene belong? (1)
(b) Give the graphic formula of an isomer of pent-1-ene. (2)
(c) Why is the bromine solution decolourised by the pent-1-ene? (1)
(d) Write down the graphic formula of the organic product. (2)
(e) What type of reaction has occurred? (1)
(f) Would you expect pentane to react with bromine solution in a similar way? Explain your answer. (2)
(g) Why do you think that butter does not decolourise the bromine solution but soft margarine does? (1)
(h) How would you expect a sample of soft margarine to react with hydrogen in the presence of a catalyst? (2)

Total 12

21.2 When butanol reacts with sodium a colourless gas is evolved.

(a) What is the colourless gas? Describe a test that you could do to confirm your answer. (2)
(b) Write an equation for the reaction of butanol with sodium and name the organic product. (2)

A similar reaction occurs between anhydrous ethanoic acid and sodium.

(c) Write an equation for the reaction. (1)
(d) Why is it necessary to use *anhydrous* ethanoic acid for this reaction? (1)

Butanol and ethanoic acid will react together to form an ester. The reaction is reversible.

(e) What name is given to this process? (1)
(f) What substance is usually added when this reaction takes place? (1)
(g) What is the function of this substance? (3)
(h) Write an equation for the reaction that takes place. (1)
(i) What conditions would encourage the backward reaction to be the predominant one? (1)
(j) Explain why these conditions favour the backward reaction. (2)

Total 15

21.3
$$\text{ethene} \longrightarrow \text{ethane}$$
$$\uparrow$$
$$\text{ethyl ethanoate} \longleftarrow \text{ethanol} \longrightarrow \text{ethanoic acid}$$
$$\downarrow \text{sodium hydroxide}$$
$$\text{sodium ethanoate}$$

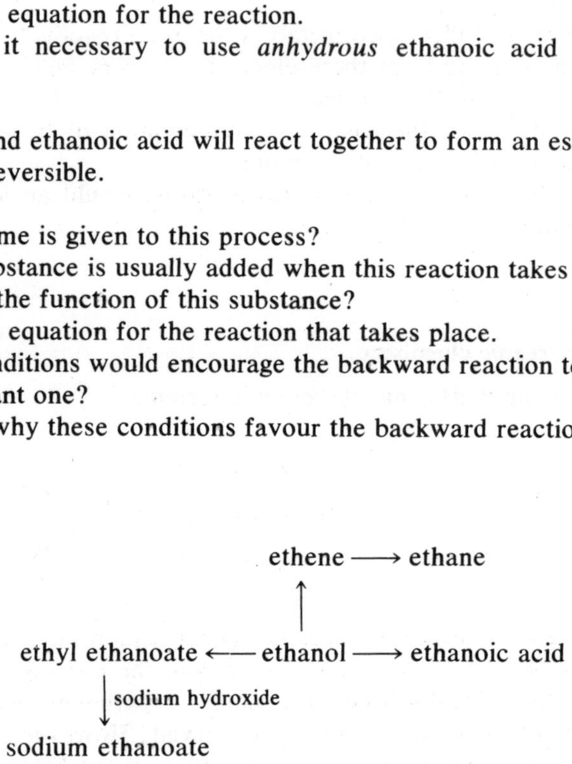

(a) Which one of the compounds shown above is ionic? (1)
(b) Which of the compounds would you expect to conduct an electric current if water were added? (2)
(c) Suggest reagents which could be used to convert ethanol to (i) ethene, (ii) ethanoic acid, (iii) ethyl ethanoate. (3)
(d) What conditions are needed to obtain ethane from ethene? (2)
(e) Write an equation for the reaction. (1)
(f) What is the other product obtained when sodium hydroxide solution is reacted with ethyl ethanoate? (1)
(g) What types of reaction occur in the changes (i) ethene to ethane, (ii) ethanol to ethene, (iii) ethanol to ethanoic acid, (iv) ethanol to ethyl ethanoate, (v) ethyl ethanoate to sodium ethanoate? (5)

Total 15

21.4 Under certain conditions ethene can be polymerised into polythene.

(a) What is meant by polymerisation? (1)
(b) What conditions are necessary for the polymerisation of ethene? (2)
(c) Write an equation showing the structural changes which occur during the reaction. (1)

If phenylethene (styrene) is heated with a little di(dodecanoyl) peroxide (lauroyl peroxide) and then poured into a beaker of ethanol, a solid separates at the bottom of the beaker. This solid softens on heating and hardens on cooling, the process being repeatable any number of times. The formula of the solid may be represented as shown:

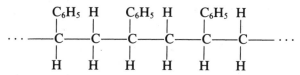

(d) What is the purpose of the di(dodecanoyl) peroxide? (1)
(e) What is the common name of the solid product? (1)
(f) Draw the formula of the repeating unit in its structure. (1)
(g) From this, draw the formula of phenylethene. (2)
(h) What type of reaction occurs during the polymerisation? (1)
(i) To which class of plastics does the product belong? (1)
(j) Name the other class of plastics. (1)
(k) How does the behaviour of the second class of plastics differ from that of the first? (1)
(l) How can the difference in behaviour be accounted for in terms of molecular structure? (2)

Total 15

## 22 Chemical analysis

22.1 A solution of potassium hydroxide (mass concentration $7.00 \text{ g dm}^{-3}$) is titrated against dilute sulphuric acid. $22.5 \text{ cm}^3$ of the acid is required to neutralise $25.0 \text{ cm}^3$ of the alkali.

(a) Name a suitable indicator for the reaction. (1)
(b) Write an equation for the reaction. (1)
D (c) What is the concentration of the alkali in mol dm$^{-3}$? (2)
(d) How many moles of potassium hydroxide are present in $25.0 \text{ cm}^3$ of solution? (2)
(e) How many moles of sulphuric acid must be present in the $22.5 \text{ cm}^3$ needed to neutralise this amount of alkali? (2)
(f) Calculate the concentration of the sulphuric acid in mol dm$^{-3}$ (2)

Total 10

22.2 A solution of sodium carbonate is titrated with dilute nitric acid, concentration 0.3 M. $25.0 \text{ cm}^3$ of the alkali requires $15.0 \text{ cm}^3$ of acid for neutralisation.

D  (a) What is the mass concentration of the acid? (2)
    (b) Write an equation for the reaction. (1)
    (c) How many moles of nitric acid are present in 15.0 cm³ of 0.3 M solution? (2)
    (d) How many moles of alkali will this neutralise? (1)
    (e) What is the concentration of the alkali in mol dm⁻³? (2)
D  (f) What mass of anhydrous sodium carbonate must be dissolved in distilled water to make 1 dm³ of this solution? (2)
    Total  10

22.3★ An acid, $H_nX$, has a relative molecular mass of 254. 25.0 cm³ of a solution of sodium hydroxide with mass concentration 3.30 g dm⁻³ requires 24.0 cm³ of a solution of $H_nX$, mass concentration 7.27 g dm⁻³, for neutralisation. The equation for the reaction is:

$$H_nX(aq) + n\,NaOH(aq) \rightarrow Na_nX(aq) + n\,H_2O(l).$$

D  (a) What is the concentration of the alkali in mol dm⁻³? (2)
    (b) How many moles of sodium hydroxide are present in 25.0 cm³ of solution? (1)
    (c) Hence, write an expression in terms of $n$ for the number of moles of $H_nX$ present in the 24.0 cm³ of acid needed to neutralise this amount of alkali. (2)
    (d) Write an expression in terms of $n$ for the number of moles of $H_nX$ in 1 dm³ of acid. (1)
    (e) Write a similar expression for the mass concentration of $H_nX$. (1)
    (f) From this, and the value for the mass concentration given in the introduction to the question, calculate the value of $n$. (2)
    (g) What is the word used to describe $n$? (1)
    Total  10

22.4★ An experiment to find the percentage of ammonia in an ammonium salt was carried out as follows.

    1.68 g of an ammonium salt was dissolved in 250 cm³ of distilled water. 25.0 cm³ of this solution was boiled with 50.0 cm³ of 0.09 M sodium hydroxide solution until all of the ammonia had been expelled from the salt. The remaining liquid required 24.0 cm³ of 0.05 M sulphuric acid for neutralisation.

    (a) Write an equation for the reaction between dilute sulphuric acid and sodium hydroxide solution. (1)
    (b) How many moles of sulphuric acid were present in the 24.0 cm³ of 0.05 M solution? (1)
    (c) How many moles of sodium hydroxide did this neutralise at the end of the experiment? (2)
    (d) How many moles of sodium hydroxide were present in the original 50.0 cm³ of 0.09 M solution? (1)
    (e) From your answers to (c) and (d), how many moles of sodium hydroxide must have been used up in expelling the ammonia from the ammonium salt? (2)

(f) The ionic equation for the reaction between an ammonium salt and sodium hydroxide solution is:

$$NH_4^+(aq) + OH^-(aq) \rightarrow H_2O(l) + NH_3(g).$$

How many moles of ammonia were expelled by the sodium hydroxide solution in the experiment? (1)

D (g) What mass of ammonia is this? (2)

(h) Hence, calculate the percentage of ammonia in the ammonium salt. (2)

Total 12

22.5★ The number of molecules of water of crystallisation, x, in hydrated barium chloride, $BaCl_2.xH_2O$, can be found from the following experimental results.

A solution was made by dissolving 3.00 g of hydrated barium chloride in 250 cm³ of water. 25.0 cm³ of this solution was boiled with 50.0 cm³ of 0.05 M sodium carbonate solution (an excess) and, after cooling, the precipitate of barium carbonate was filtered off. The filtrate (sodium chloride solution and excess sodium carbonate solution) required 25.4 cm³ of 0.1 M hydrochloric acid for neutralisation.

(a) Write an equation for the reaction between the sodium carbonate in the filtrate and the 0.1 M hydrochloric acid. (1)
(b) How many moles of hydrogen chloride were present in the 25.4 cm³ of 0.1 M hydrochloric acid? (1)
(c) How many moles of sodium carbonate must have been present in the filtrate? (2)
(d) How many moles of sodium carbonate were present in the original 50.0 cm³ of 0.05 M solution? (1)
(e) Hence, how many moles of sodium carbonate reacted with the barium chloride? (1)
(f) Write an equation for the reaction between barium chloride solution and sodium carbonate solution. (1)
(g) How many moles of barium chloride were present in the 25.0 cm³ of solution used? (2)
(h) In terms of the molar mass, $M$, of $BaCl_2.xH_2O$, how many grams of the hydrated salt is this? (2)
(i) From the figures given in the introduction to the question and your answer to (h), calculate the value of $M$. (2)

D (j) Hence, find the value of x. (2)

Total 15

22.6 A metal ore has the name 'cerussite'. Chemical analysis is carried out and the following observations are made.

The ore is insoluble in water. On being heated it turns from white to yellow and gives off a colourless gas which turns lime water milky.

(a) Which anion does the ore contain? (2)
(b) Describe one other test that could be carried out on the original substance to confirm your answer. (2)

(c) Which anion is probably present in the yellow solid? (2)

When dilute nitric acid is added to the yellow solid it dissolves, forming a colourless solution. The addition of sodium hydroxide solution also dissolves the yellow solid and produces a colourless solution.

(d) Which word is used to describe substances that react with both acids and alkalis? (1)

Addition of dilute hydrochloric acid to the solution of the yellow solid in nitric acid produces a white precipitate. This is filtered off and found to dissolve in hot water; as the solution cools, the solid is redeposited as crystals.

(e) Suggest what the crystals might be. (1)
(f) Hence, name the chemical in 'cerussite'. (2)
(g) Write equations for
    (i) the thermal decomposition of 'cerussite', (1)
    (ii) the action of dilute nitric acid on the residue, (1)
    (iii) the reaction of dilute hydrochloric acid with the solution obtained in (ii). (1)
(h) Describe what would happen if potassium iodide solution were added to the solution obtained in (g) (ii). (1)
(i) Write an ionic equation for the reaction. (1)

Total 15

22.7★ A, B, C and D are four aqueous solutions with pH values of 7, 12, 1 and 14 respectively. They are mixed together in the exact proportions required for reaction and the results are as shown in the table.

|        | A | B | C |
|---|---|---|---|
| With B | White precipitate, neutral solution. | | |
| With C | White precipitate, acidic solution. | Effervesces, neutral solution. | |
| With D | White precipitate, neutral solution. | No reaction, alkaline solution. | No *visible* reaction, neutral solution. |

(a) What type of substance is solution C? (1)
(b) Name three anions that could react with C to produce a gas. (3)
(c) Suggest which anion B contains. Give a reason for your answer. (2)
(d) What do you think D is? How did you reach this conclusion? (3)
(e) The white precipitate obtained when A reacts with D is soluble in excess alkali. Name three cations that this precipitate might contain. (3)

(f) The precipitate obtained when A reacts with C is soluble in hot water but reappears on cooling. What do you think this precipitate is? (1)
(g) Which cation does A contain? (1)
(h) Hence suggest what solution A might be. (2)
(i) Name the precipitate obtained when A reacts with B. (1)
(j) What do you think solution C is? Give reasons for your answer. (3)

Total 20

# Part II

1 (5, 11, 22)

Platinum electrodes were placed in a beaker containing 50 cm³ of barium hydroxide solution and an electric current was passed between them. 1 M sulphuric acid was added in 2.0 cm³ portions and the current was noted after each addition. The results which were obtained are shown in the table.

| Volume of $H_2SO_4$/cm³ | 0.0 | 2.0 | 4.0 | 6.0 | 8.0 | 10.0 |
|---|---|---|---|---|---|---|
| Current/mA | 50 | 30 | 10 | 14 | 42 | 70 |

(a) What would be seen in the beaker as the acid was added? (1)
(b) Write a chemical equation for the reaction. (1)
(c) Plot a graph of current/mA ($y$ axis) against volume of acid/cm³ ($x$ axis). (3)
(d) Why did the current drop as the acid was added? (1)
(e) What was the value of the current at the end-point of the titration? (1)
(f) Why was this particular value obtained? (1)
(g) What volume of acid had been added at the end-point? (1)
(h) What was the concentration of the barium hydroxide solution in mol dm⁻³? (2)
(i) What volume of acid would be needed to reach the end-point if the barium hydroxide solution were replaced by sodium hydroxide solution with the same concentration (in mol dm⁻³)? (2)
(j) Would the current drop to the same value? Give a reason for your answer. (2)

Total 15

2 (5, 13)

The table below shows the time taken for the same mass of magnesium to dissolve completely in sulphuric acid of various concentrations.

| Concentration/mol dm⁻³ | ½ | 1 | 2 | 4 | 8 | 12 | 18 |
|---|---|---|---|---|---|---|---|
| Time/s | 450 | 45 | 22 | 5 | 106 | 750 | very little reaction |

(a) Write (i) the chemical equation (ii) the ionic equation for the reaction between magnesium and sulphuric acid. (2)
(b) What happens to the rate of the reaction as the concentration of the acid increases from $\frac{1}{2}$ M to 4 M? (1)
(c) Explain why the rate changes in this way. (1)
(d) Would you expect to see any reaction between magnesium and a solution of hydrogen chloride in (i) water (ii) methylbenzene? (2)
(e) Explain your answer to (d). (4)
(f) Do you think that an 18 M solution of sulphuric acid contains many ions? Give a reason for your answer. (2)
(g) From the data given, how does the rate of the reaction change when the acid becomes more concentrated than 4 M? (1)
(h) Suggest why the rate should change in this way. (1)
(i) Which solution of sulphuric acid would be the best conductor of electricity? (1)

Total 15

3 (6, 11)

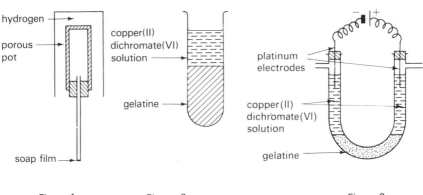

Figure 1        Figure 2        Figure 3

Figure 1 illustrates an experiment in which hydrogen surrounds a porous pot containing air. After some time it is noticed that a soap bubble is blown at the lower end of the tube.

(a) How does the pressure of the gas inside the porous pot compare with atmospheric pressure? (1)
(b) Explain why the pressure varies in this way. (1)
(c) How would you expect the results to differ if carbon dioxide were used instead of hydrogen? (1)

In a second experiment, an aqueous solution of gelatine is allowed to set in a boiling tube and a solution of copper(II) dichromate(VI) is poured on top (Figure 2). After some time, a green colour due to the copper(II) dichromate(VI) is seen in the gelatine layer.

(d) Explain how the copper(II) dichromate(VI) gets into the gelatine layer. (2)

The copper(II) dichromate(VI) solution is poured away and replaced by distilled water.

(e) What would you expect to see several hours after doing this? (2)
(f) What name is used to explain the phenomenon illustrated by these two experiments? (1)
(g) Would you expect noticeable results to be obtained more quickly in experiment 1 or experiment 2? Explain your answer. (2)

Finally, the apparatus illustrated in Figure 3 is assembled. After some minutes, a blue colour is observed in the gelatine near the positive electrode whilst an orange colour is noticed in the gelatine near the negative electrode.

(h) Which ions are present in copper(II) dichromate(VI)? (1)
(i) What colour is associated with each of these ions? (2)
(j) Hence, explain why the colours seen in the gelatine are in the positions described. (2)

Total 15

4 (6, 7, 8)

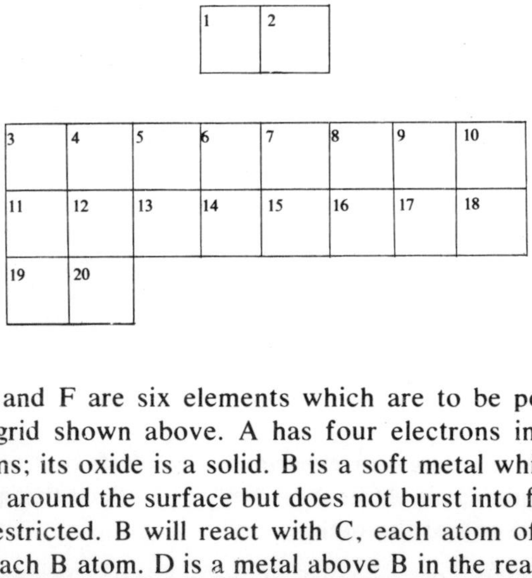

A, B, C, D, E and F are six elements which are to be positioned in the periodic table grid shown above. A has four electrons in the outermost shell of its atoms; its oxide is a solid. B is a soft metal which when added to water shoots around the surface but does not burst into flames unless its movement is restricted. B will react with C, each atom of C gaining one electron from each B atom. D is a metal above B in the reactivity series. E is in group VII of the periodic table and is liberated from a solution of DE when C is bubbled through the solution. F is a colourless gas with one electron in the outermost shell of its atoms.

Use the letters A–F in answering the following questions: do not use chemical symbols.

(a) Copy out the grid and write the six letters A–F in their correct positions. (6)
(b) What type of bonding would you expect when (i) A combines with E, (ii) C combines with D? (2)
(c) Draw diagrams of the electronic structures of the compounds formed in (b). (4)

(d) State whether you would expect each compound in (b) to have a high or a low melting point and give a reason for your answer. (4)
(e) Would a liquid sample of each compound conduct electricity? Explain your answer. (2)
(f) What would you expect to happen if the compounds were placed separately in water? (2)

Total  20

5 (9, 20)

A white crystalline solid is known to be a compound of lead and bromine. 0.734 g of the compound is added to distilled water, together with 0.05 g of aluminium powder (an excess), and the mixture boiled for a short time. Excess sodium hydroxide solution is added, causing an effervescence, and the gas evolved pops when a flame is applied. A bead of lead remains at the bottom of the liquid and this is removed, washed with propanone and left in a warm oven for about half an hour. The mass of the bead is then found to be 0.414 g.

(a) Why does aluminium displace lead from the lead compound? (1)
(b) Which gas is evolved when the sodium hydroxide solution is added? (1)
(c) What is the purpose of the sodium hydroxide solution? (1)
(d) Why is it important to have an excess of aluminium powder? (1)
(e) What is the purpose of the propanone? (1)
D (f) How many moles of lead atoms are present in the bead of lead? (2)
(g) What is the mass of bromine in the original compound? (1)
D (h) How many moles of bromine atoms is this? (2)
(i) From your answers to (f) and (h), calculate the empirical formula of the lead compound. (2)

Total  12

6 (9, 12, 18, 20)

When 8.34 g of green hydrated iron(II) sulphate is gently heated, water of crystallisation is driven off and 4.56 g of the white anhydrous form of the salt remains. Strong heating of this yields a reddish-brown solid and the vapour given off turns a moist universal indicator paper red.

Treatment of the hydrated salt with concentrated sulphuric acid also produces a white solid.

(a) Explain what is meant by (i) water of crystallisation, (ii) hydrated, (iii) anhydrous. (3)
(b) What mass of water of crystallisation is contained in the sample of hydrated iron(II) sulphate? (1)
D (c) Calculate the empirical formula of hydrated iron(II) sulphate. (2)
(d) Write an equation to show what happens when the hydrated salt is gently heated. (1)
(e) Which gas or gases is/are responsible for causing the universal indicator to turn red? (2)

(f) Is a reddish-brown colour generally associated with iron(II) or iron(III) compounds? (1)
(g) What do you think the reddish-brown solid is? (1)
(h) What is the white solid obtained when the hydrated salt is treated with concentrated sulphuric acid? (1)
(i) What is the term used to describe the action of the sulphuric acid? (1)
(j) Give a definition of this term. (1)
(k) Explain how the action of the sulphuric acid in this experiment differs from the action of the acid when it is used to dry gases. (1)

Total 15

7 (9, 10, 16, 21)

Carbon and silicon (Si) are both in group IV of the periodic table.

Magnesium burns in carbon dioxide to produce a white ash of magnesium oxide and black specks of carbon.

A solid compound of magnesium, magnesium silicide, has the following composition by mass: Mg = 63.2%, Si = 36.8%. This compound reacts with dilute hydrochloric acid to give magnesium chloride solution and a mixture of gases, one of which, silane, has composition by mass Si = 87.5%, H = 12.5%, and relative molecular mass of 32.

(a) Write an equation for the reaction between magnesium and carbon dioxide. (1)
(b) What would you expect to be formed if magnesium and silicon(IV) oxide, $SiO_2$, were heated together? (2)
D (c) Calculate the empirical formula of magnesium silicide. (2)
D (d) Calculate the empirical formula of silane. (2)
D (e) What is the molecular formula of this gas? (1)
(f) Write an equation for the formation of silane from magnesium silicide and dilute hydrochloric acid. (2)
D (g) What volume of silane, measured at s.t.p., would be obtained from 4 g of magnesium silicide? (2)
(h) Which compound of carbon does silane resemble? (1)
(i) What would you expect to be formed when silane is burned in air? (2)

Total 15

8 (4, 9, 17, 20)

When a blue crystalline solid, A, is heated gently in a test tube, brown fumes of a gas, B, are seen and a black solid, C, remains in the tube.

(a) Name A, B and C. (3)
(b) A colourless gas is also obtained. Suggest what this might be and describe an experiment that you could do which would confirm your answer. (2)
(c) Write an equation for the thermal decomposition of A. (1)

When dry hydrogen is passed over 2 g of C in a suitable apparatus, a reddish-brown solid, D, is obtained.

(d) Draw a diagram of the apparatus used for the reaction between hydrogen and C. (3)
(e) Name D. There is another product of the reaction. Give two simple tests that would confirm the presence of this product. (3)
(f) Dry ammonia can also be used to convert C into D. Give an equation for the reaction that occurs. (1)
D (g) How many moles is 2 g of C? (2)
(h) Hence, how many moles of D should be produced? (1)
D (i) What is the expected yield of D in grams? (2)
(j) 1.55 g of D is obtained. Calculate the actual percentage yield. (2)

Total 20

9 (5, 9, 17, 20)

A farmer had his soil tested. The report stated that the soil had a pH value of 5 and was lacking in nitrogen. He was advised to apply lime and a nitrogenous fertiliser.

(a) Why did the soil need lime? (1)
(b) Should the farmer apply quicklime (calcium oxide) or slaked lime (calcium hydroxide)? Give a reason for your answer. (2)
(c) What is meant by a 'nitrogenous fertiliser'? (1)
D (d) Two substances which could be used are ammonium sulphate and ammonium nitrate. Calculate the percentage of nitrogen in each one. (4)
(e) If the cost of the ammonium sulphate is two-thirds that of the ammonium nitrate, which substance is the cheaper source of nitrogen? Show how you reached your answer. (3)
(f) What must happen to the ammonium ions before the nitrogen in them can be taken in by plants? (1)
(g) What in the soil brings about this change? (1)
(h) Should the farmer apply lime and ammonium salts together, or at different times? Explain your answer. (2)

Total 15

10 (9, 14, 18)

In the contact process for the manufacture of sulphuric acid sulphur is burned to sulphur dioxide and this is combined with oxygen in the presence of a complex vanadium catalyst to form sulphur trioxide. The equation for the reaction is

$$2SO_2(g) + O_2(g) \rightleftharpoons 2SO_3(g) \qquad \Delta H - ve.$$

(a) Would you expect the best yield to be obtained at a high or at a low pressure? Explain your answer. (3)
(b) In practice the reaction is carried out at atmospheric pressure. Why is this? (1)
(c) Would a high yield be favoured by a high or a low temperature? Explain your answer. (3)

(d) The process usually operates at about 720 K (450°C). Why is this temperature chosen? (2)
(e) What is the purpose of the catalyst? (1)

Although sulphur trioxide combines with water to form sulphuric acid, in this process it is absorbed in concentrated sulphuric acid and the resulting product diluted with water according to the following equations:

$$H_2SO_4(l) + SO_3(g) \rightarrow H_2S_2O_7(l)$$
$$H_2S_2O_7(l) + H_2O(l) \rightarrow 2H_2SO_4(l)$$

(f) Why is this procedure adopted? (1)
D (g) What is the maximum mass of sulphuric acid which could be obtained from 3200 kg of sulphur? (2)
(h) Why would this yield never be obtained? (2)
Total 15

11 (16, 18)

When sodium carbonate is reacted with dilute hydrochloric acid a gas, X, is evolved. When sodium sulphite is similarly treated with acid a gas, Y, is produced.

(a) Name X and Y. (2)
(b) Give two physical properties which are common to both X and Y. (2)
(c) Describe what happens when strips of burning magnesium are lowered into separate gas jars of X and Y. Give an equation in each case and name any solid products. (5)
(d) When X is bubbled into an acidified solution of potassium manganate(VII) no reaction occurs. Similar treatment of Y leads to decolourisation of the solution. Explain why this is so. (1)
(e) What would you observe if X and Y were bubbled separately into different samples of barium chloride solution? Explain each reaction. (3)
(f) How would the result differ if the barium chloride solution were first acidified with dilute hydrochloric acid? (2)
Total 15

12 (7, 11, 19)

Fluorine, F, is at the top of group VII in the periodic table. It is prepared by the electrolysis of liquid hydrogen fluoride containing a little potassium fluoride to increase the electrical conductivity.

(a) At which electrode will the fluoride ions be discharged? (1)
(b) Write an equation to show what is happening at this electrode. (1)
(c) Which ions will be attracted to the other electrode? (1)
(d) Which of these ions will be discharged? Why did you choose this answer? (2)
(e) How does the chemical reactivity of fluorine compare with that of other non-metals? (1)

Astatine, At, comes at the foot of group VII in the periodic table.

(f) Which of the halogens, chlorine, bromine or iodine, does astatine most closely resemble? Give a reason for your answer. (1)
(g) How soluble would you expect astatine to be in water? Will it react with water to an appreciable extent? (2)
(h) There is some evidence for the existence of the At⁺ ion. What can you say about the metallic nature of the halogens as the group is descended? (1)

Total 10

13 (1, 9, 19)

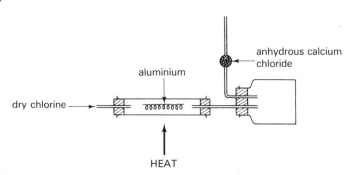

The apparatus shown above is used to prepare anhydrous aluminium chloride. The product is collected as a white solid in the bottle and can be shown to contain 20.2% by mass of aluminium and to have a relative molecular mass of 267.

When anhydrous aluminium chloride is boiled with water, the vapour evolved turns universal indicator paper red and when tested with a drop of silver nitrate solution on a glass rod, forms a white precipitate in the drop.

D (a) Calculate the empirical formula of aluminium chloride. (3)
D (b) What is its molecular formula? (2)
(c) Write an equation for the reaction which takes place in the combustion tube. (1)
(d) By what process does the aluminium chloride get from the combustion tube to the flask? (1)
(e) Give a definition of this process. (1)
(f) What is the function of the anhydrous calcium chloride? (1)
(g) Name the white precipitate formed in the silver nitrate solution. (1)
(h) Hence, name the vapour evolved when aluminium chloride reacts with water. (1)
(i) What is the other product of the reaction in (h)? (1)

Total 12

14 (9, 11, 20)

Three metallic elements, A, B and C, have the following properties.

| A | B | C |
|---|---|---|
| Extracted by chemical reduction of oxide | Extracted by electrolysis of molten chloride | Occurs as free element in nature |
| Most stable oxidation state in compounds = +3 | Oxidation state in compounds = +1 | Most stable oxidation state in compounds = +1 |

(a) Give the formula of an oxide of A. (1)
(b) Suggest a reducing agent that might be used industrially to extract A from its oxide. (1)
(c) Give the formula of the chloride of B. (1)
(d) Write equations to illustrate what is happening at the electrodes during the electrolysis of the molten chloride of B. (2)
D (e) 85 g of the chloride of B will yield a maximum of 71 g of chlorine during electrolysis. Calculate the relative atomic mass of B. (3)
D (f) Which element do you think B is? (1)
(g) In which order do you think A, B and C come in the reactivity series? (2)
(h) What would you expect to observe if a sample of metal A were placed in a solution of the nitrate of C? Explain your answer. (2)
(i) Ions of B and C are present in solution together. Which ones will be preferentially discharged during electrolysis? Explain how you obtained your answer. (2)
(j) What would you expect to happen if a sample of B were added to water? Give the approximate pH value of any solution obtained. (3)
(k) Would you expect to observe any reaction between C and water? What pH would be registered this time? (2)

Total 20

15 (5, 7, 8, 20)

The metal strontium (Sr) comes just below calcium in the periodic table and just above calcium in the reactivity series.

(a) How many electrons does an atom of strontium have in its outermost shell? (1)
(b) Write the formula for the ion that strontium forms in its compounds. (1)
(c) What would you expect to happen if a lump of strontium were added to water? (2)
(d) Write an equation for the reaction. (1)
(e) Would you expect the pH of the solution to be between 1 and 6, 7, or between 8 and 14? (1)
(f) If carbon dioxide is bubbled into the solution obtained in (c), a white precipitate is formed. Name this precipitate and write an equation for the reaction. (2)

(g) How would you expect strontium to react with dilute hydrochloric acid? Name the products of this reaction. (3)
(h) What would happen if (i) sodium nitrate solution, (ii) sodium sulphate solution were added to the mixture obtained in (g)? (2)

Strontium chloride has a high melting point and will conduct electricity when molten.

(i) What type of structure would you expect strontium chloride to have? (2)
(j) Outline briefly how this structure accounts for the physical properties mentioned above. (3)
(k) Would you expect strontium chloride to be soluble in water? Give a reason for your answer. (2)

Total 20

16★ (9, 10, 21)

A solid consists of 57% of magnesium and 43% of carbon by mass. Its relative molecular mass is 84. 0.21 g of the solid reacts with water to form magnesium hydroxide, together with 60 cm³ of a gaseous hydrocarbon $C_xH_y$.

The 60 cm³ of hydrocarbon requires 240 cm³ of oxygen for complete combustion and produces 180 cm³ of carbon dioxide (all volumes are measured at room temperature and pressure).

(a) What is formed in addition to carbon dioxide when a hydrocarbon burns in a plentiful supply of oxygen? (1)
(b) Work out an equation in terms of x and y for the combustion of $C_xH_y(g)$. (2)
(c) What is the ratio of moles of $C_xH_y(g):O_2(g):CO_2(g)$ in the experiment? (2)
(d) Using your answers to (b) and (c), work out the molecular formula of the hydrocarbon. (3)
(e) Draw a graphic structure for the hydrocarbon. (2)
(f) Is the hydrocarbon saturated or unsaturated? Explain your answer. (2)
(g) What would you expect to see if the hydrocarbon were bubbled into acidified potassium manganate(VII) solution? (1)
D (h) Find the empirical formula of the original solid and suggest a name for it. (3)
D (i) What is the ratio of moles of the original solid : moles of $C_xH_y(g)$? (2)
(j) Work out the equation for the reaction between the solid and water. (2)

Total 20

17 (9, 10, 21)

A substance X contains 92.3% of carbon and 7.7% of hydrogen by mass and has a relative molecular mass of 26.

D (a) Calculate the empirical formula of X. (2)
D (b) What is its molecular formula? (1)

X can be polymerised in two different ways to give Y or Z.

(c) Give the empirical formula of Y and Z. (1)

Y is a yellow liquid. A molecule of Y can react with one molecule of bromine to give a single product P of relative molecular mass 264.

(d) What general name is given to a reaction in which two or more substances combine together to give a single product? (1)
(e) Give one other example of a reaction of this type and name the starting materials and the product. (2)
(f) How many bromine atoms are present in one molecule of P? (1)
D (g) Hence, calculate the relative molecular mass of Y. (1)
(h) What is the molecular formula of Y? Write an equation to show its formation from X. (2)

Z is a colourless liquid which reacts with bromine by a substitution reaction.

(i) Explain what is meant by a substitution reaction and give one other example, naming the reactants and products. (4)

For the complete combustion of $10 \text{ cm}^3$ of Z vapour, $75 \text{ cm}^3$ of oxygen are required. $60 \text{ cm}^3$ of carbon dioxide and $30 \text{ cm}^3$ of steam are produced, all volumes being measured at the same temperature and pressure.

(j) What is the ratio of volumes of $Z(g):O_2(g):CO_2(g):H_2O(g)$? (1)
(k) What is the ratio of moles of $Z(g):O_2(g):CO_2(g):H_2O(g)$? (1)

The equation is of the form

$$aZ(g) + bO_2(g) \rightarrow cCO_2(g) + dH_2O(g).$$

(l) What are the values of a, b, c and d? (1)
(m) Hence, calculate the numbers of carbon and hydrogen atoms in one molecule of Z and write the formula of Z. (2)

Total 20

18★ (9, 21)

A liquid hydrocarbon, A, contains 87.8% of carbon by mass and has a relative molecular mass of 82. Its density is $0.81 \text{ g cm}^{-3}$. $1.00 \text{ cm}^3$ of A reacts rapidly with 1.58 g of bromine to form a new compound, B.

(a) Which element or elements are present in A in addition to carbon? (1)
(b) What is/are the percentage(s) by mass? (1)
D (c) Calculate the empirical formula of A. (2)
D (d) What is the molecular formula of A? (1)
(e) Does A appear to be a saturated or an unsaturated compound? (1)
(f) What type of reaction takes place between A and bromine? (1)
(g) How many moles of A molecules are present in $1.00 \text{ cm}^3$? (1)

D  (h) How many moles of bromine molecules are present in 1.58 g? (2)
    (i) How many bromine molecules would react with each double bond in A? (1)
    (j) How many double bonds does one molecule of A contain? (1)
    (k) Draw the graphic formula of A. (2)
    (l) Draw the graphic formula of B. (1)
                                        Total  15

19★ (10, 16, 21)

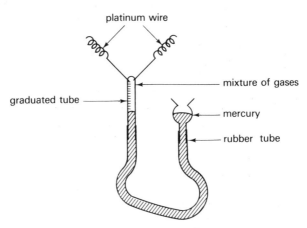

$9 \text{ cm}^3$ of a gaseous hydrocarbon and $80 \text{ cm}^3$ of oxygen (both measured at room temperature and pressure) were placed in the apparatus shown. An electric spark was passed between the platinum wires to ignite the mixture. After allowing the products to cool to room temperature the right-hand tube was raised until the two mercury surfaces were level. The volume of gas in the graduated tube was found to be $62 \text{ cm}^3$. Some concentrated potassium hydroxide solution was introduced into the graduated tube and, after the mercury surfaces had been levelled again, the volume of gas was $26 \text{ cm}^3$.

(a) Why were the mercury surfaces levelled before reading the gas volumes? (1)
(b) What was the purpose of the potassium hydroxide solution? (1)
(c) Name the gas left at the end of the experiment. (1)
(d) If the formula of the hydrocarbon is written as $C_xH_y$, the equation is of the form
$$C_xH_y(g) + ?O_2(g) \rightarrow xCO_2(g) + \tfrac{y}{2}H_2O(l).$$
Work out the number of oxygen molecules required in terms of x and y. (1)
(e) What volume of carbon dioxide was produced in the experiment? (1)
(f) What volume of oxygen was used up during the experiment? (1)
(g) From the equation, what is the ratio of moles of $C_xH_y(g):O_2(g):CO_2(g)$? (1)
(h) From the experiment, what is the ratio of moles of $C_xH_y(g):O_2(g):CO_2(g)$? (2)

(i) From your answers to (g) and (h), what is the value of x? (1)
(j) What is the value of y? (2)
(k) Draw a possible graphic formula for $C_xH_y$ and give its name. (3)

Total 15

20 (4, 18, 19)

The information given below concerns the reaction of hydrochloric acid with three black solids, A, B and C.

When A reacts with the acid, a gas D is obtained. This gas pops when a flame is applied. The reaction of B with the acid gives a greenish-yellow gas E which bleaches a moist universal indicator paper. C reacts with hydrochloric acid to give a gas F, smelling of bad eggs. D reacts with E to give a colourless gas G which fumes in moist air and gives dense white fumes if a glass rod dipped in ammonia solution is brought near.

(a) What is gas D? (1)
(b) Which type of substance will give D when it reacts with dilute hydrochloric acid? (2)
(c) What is the gas E? (1)
(d) Hence suggest what B might be. (2)
(e) What do you think F is? (1)
(f) Hence suggest a name for solid C. (2)
(g) Write an equation to show how C reacts with the acid. (1)
(h) Name the gas G. (1)
(i) Write an equation to show how D reacts with E. (1)
(j) What are the white fumes obtained when G reacts with ammonia solution? (1)
(k) What would you expect to see if silver nitrate solution were used instead of the ammonia solution? Name the products of the reaction. (2)

Total 15

21 (17, 19, 20, 21)

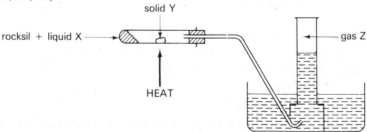

The apparatus shown may be used to prepare various gases.

(a) If Z is ethene, what are X and Y? (2)
(b) If Z is a mixture of hydrocarbons, what are X and Y? (2)
(c) If Z is hydrogen, what are X and Y? (2)
(d) If Z is oxygen, what are X and Y? (2)
(e) What is the purpose of the rocksil? (1)

(f) Why is the rocksil not heated? (1)
(g) What precaution should be taken before heating is stopped? (1)
(h) Why is this precaution taken? (1)
(i) Would this method be suitable for preparing a liquid? Explain your answer. (1)
(j) Why could Z never be hydrogen chloride? (2)

Total 15

22 (6, 16, 21)

Diamond and graphite are *allotropes*; chlorine-35 and chlorine-37 are *isotopes*; propan-1-ol and propan-2-ol are *isomers*.

(a) Explain the meanings of the words in italics. (3)

Diamond and graphite both have high melting points but diamond is hard and graphite is soft. Both consist entirely of carbon atoms.

(b) What general type of structure do both possess? (1)
(c) Draw a diagram of the structure of each one. (2)
(d) Explain how the structures account for the difference in hardness of diamond and graphite. (2)
(e) From the structures you have given, predict, giving reasons, which of the two would have the lower density. (2)

The symbols for chlorine-35 and chlorine-37 are $^{35}_{17}Cl$ and $^{37}_{17}Cl$ respectively. The isotopes both have the same chemical properties.

(f) How many protons, neutrons and electrons are present in the two types of atom? (2)
(g) Give the electronic configuration of each isotope. (1)
(h) Why do both isotopes have the same chemical properties? (1)
(i) Explain why the relative atomic mass of chlorine is 35.5. (1)

The structures of propan-1-ol and propan-2-ol are:

```
   H  H  H                           H  H      H
   |  |  |                           |  |      |
H—C—C—C—O—H         and         H—C—C————C—H
   |  |  |                           |  |      |
   H  H  H                           H  O—H    H
   propan-1-ol                       propan-2-ol
```

(j) To which homologous series do the two compounds belong? (1)
(k) Which gas would be given off if sodium were added to propan-1-ol? Would propan-2-ol give this gas? (2)
(l) What type of compound would be formed if either of the compounds reacted with ethanoic acid? (1)
(m) Draw the structure of an isomer of propan-1-ol and propan-2-ol which does not belong to the same homologous series. (It does not matter if you do not know what the substance is called.) (1)

Total 20

# Answers to calculations

**Part I**

1.2 (b) (i) 0.5, (ii) 0.4, (iii) 0.2, (iv) 0.7
 (e) 0.5, 0.4, 0.2

3.2 (c) 13 cm$^3$
 (d) 13%
 (f) 21%

3.3 (f) 20%

6.1 (c) 0.05 cm$^3$
 (d) 0.0001 cm$^3$
 (e) 0.000005 cm$^3$
 (f) 0.000015 cm$^3$
 (g) 154 cm$^2$
 (h) 0.0000001 cm

6.3 (a) 19
 (b) 20
 (c) 2.8.8.1
 (d) 10
 (e) 10
 (f) 2.8
 (g) 6
 (h) 12
 (i) 2.4
 (j) 6
 (k) 14
 (l) 6

6.4 (b) (i) 19, (ii) 20, 21, 22, (iii) 19
 (f) 39.14
 (j) (i) 1.4 × 10$^9$ years, (ii) 2.8 × 10$^9$ years

9.1 (a) 0.32
 (b) 0.02
 (c) (i) 4.14, (ii) 0.02
 (d) PbO
 (e) Pb$_3$O$_4$
 (f) 0.003
 (g) 0.006
 (h) 2

9.2 (c) 0.13 g
 (d) 0.002
 (f) 0.5 g
 (g) 0.0039
 (h) ZnI$_2$

9.3 (c) (i) 3.0 cm, (ii) 6.0 cm$^3$
 (d) 0.003
 (e) 0.0045
 (f) 2:3
 (h) 16.0 cm$^3$

 (i) 0.012
 (j) 4

10.1 (b) 0.002
 (c) 42
 (e) 14.3%
 (f) CH$_2$
 (g) C$_3$H$_6$

10.2 (a) 1:2:2
 (b) 1:2:2
 (d) 0.08, 0.08
 (e) 0.06
 (f) 4:3:4

10.3 (a) Al$_4$C$_3$
 (c) 1:3
 (e) 1:2:1
 (f) 1:2:1

11.3 (f) 25.25 g
 (h) 27.75 g
 (i) 2.5 g
 (j) 0.023
 (k) 2250
 (l) 0.023

11.4 (b) 0.00419
 (c) 0.00419
 (d) 405
 (e) 96 600 C
 (f) 96 600 C mol$^{-1}$
 (h) 0.00419
 (i) 0.00210
 (j) 50.4 cm$^3$

11.5 (b) 1930
 (c) 0.02
 (d) 2.16 g
 (e) 0.02
 (g) 0.59 g
 (h) 0.01
 (i) +2

11.6 (e) 0.0015
 (f) 0.006
 (g) 579
 (h) 0.104 g
 (i) 0.002
 (j) 3

12.2 (a) (i) 0.2 mol dm$^{-3}$, (ii) 0.4 mol dm$^{-3}$

13.2 (c) 0.085 g

15.1 (a) −5976 kJ mol$^{-1}$

15.2 (b) 0.02
 (c) 0.01
 (d) 0.0126
 (e) 3.5°C
 (f) 1470 J
 (g) $-147$ kJ (mol Cu)$^{-1}$
 (i) 2
 (j) 193 000 C
 (k) 88 800 J
15.3 (a) 567 J
 (b) 0.01
 (c) $-56.7$ kJ mol$^{-1}$
 (d) $-105$ J
 (e) 0.01
 (f) 10.5 kJ mol$^{-1}$
 (h) $-67.2$ kJ mol$^{-1}$
20.3 (d) 25% temporary, 75% permanent
 (e) 0.0001
 (f) 8.0
 (g) 0.0000125
 (h) 0.00015
22.1 (c) 0.125 mol dm$^{-3}$
 (d) 0.00313
 (e) 0.00157
 (f) 0.0698 mol dm$^{-3}$
22.2 (a) 18.9 g dm$^{-3}$
 (c) 0.0045
 (d) 0.00225
 (e) 0.09 mol dm$^{-3}$
 (f) 9.54 g
22.3 (a) 0.0825 mol dm$^{-3}$
 (b) 0.00206
 (c) $\dfrac{0.00206}{n}$
 (d) $\dfrac{0.0858}{n}$
 (e) $\dfrac{21.8}{n}$ g dm$^{-3}$
 (f) 3.00
22.4 (b) 0.0012
 (c) 0.0024
 (d) 0.0045
 (e) 0.0021
 (f) 0.0021
 (g) 0.0357 g
 (h) 21.25%
22.5 (b) 0.00254
 (c) 0.00127
 (d) 0.0025
 (e) 0.00123
 (g) 0.00123
 (h) $0.00123 \times M$

 (i) 244
 (j) 2

**Part II**

1 (e) 0 mA
 (g) 5 cm$^3$
 (h) 0.1 mol dm$^{-3}$
 (i) 2.5 cm$^3$
5 (f) 0.002
 (g) 0.32 g
 (h) 0.004
 (i) PbBr$_2$
6 (b) 3.78 g
 (c) FeSO$_4$.7H$_2$O
7 (c) Mg$_2$Si
 (d) SiH$_4$
 (e) SiH$_4$
 (g) 1.18 dm$^3$
8 (g) 0.0252
 (h) 0.0252
 (i) 1.60 g
 (j) 96.9%
9 (d) (NH$_4$)$_2$SO$_4$: N = 21.2%
   NH$_4$NO$_3$: N = 35%
 (e) NH$_4$NO$_3$
10 (g) 9800 kg
13 (a) AlCl$_3$
 (b) Al$_2$Cl$_6$
14 (e) 7
16 (c) 1:4:3
 (d) C$_3$H$_4$
 (h) Mg$_2$C$_3$
 (i) 1:1
17 (a) CH
 (b) C$_2$H$_2$
 (g) 104
 (h) C$_8$H$_8$
 (j) 2:15:12:6
 (k) 2:15:12:6
 (m) C$_6$H$_6$
18 (b) 12.2%
 (c) C$_3$H$_5$
 (d) C$_6$H$_{10}$
 (g) 0.00988
 (h) 0.00988
19 (d) $(x + \frac{y}{4})$
 (e) 36 cm$^3$
 (f) 54 cm$^3$
 (g) $1 : (x + \frac{y}{4}) : x$
 (h) 1:6:4
 (i) 4
 (j) 8